Stories from the
Lobster Fishery
of
Cumberland's Northern Shore

Stephen Leahey

North Cumberland Historical Society
Pugwash, Nova Scotia

Lockeport, Nova Scotia

Design: Brenda Conroy
Cover photo: workers at the Saddle Island lobster cannery,
circa 1910, courtesy of Burt Langille
Printed and bound in Canada by
Hignell Printing, Winnipeg, MB

Published by Community Books
RR1, Lockeport, Nova Scotia, B0T 1L0
phone/fax: (902) 656-2223
email: kathleen.tudor@ns.sympatico.ca
www.selfpublishingspecialists.com
and
North Cumberland Historical Society
P.O. Box 353, Pugwash, NS B0K 1L0
nchs_2@yahoo.ca
(Please contact the North Cumberland Historical Society
to obtain copies of this book.)

Library and Archives Canada Cataloguing in Publication

Leahey, Stephen Gerard
Stories from the lobster fishery of Cumberland's northere shore/
Stephen Gerard Leahey

ISBN 1-896496-49-0

1. Lobster fisheries—Nova Scotia—Cumberland (County)—
Anecdotes. 2. Lobster fishers—Nova Scotia—Cumberland
(Conty)—Anecdotes. I. Title.

SH380.2. C3L42 2005   639.54   C2005-902244-2

To my mother, Greta Leahey-Dow,
a dedicated member of the North Cumberland Historical Society
and the inspiration behind this book.

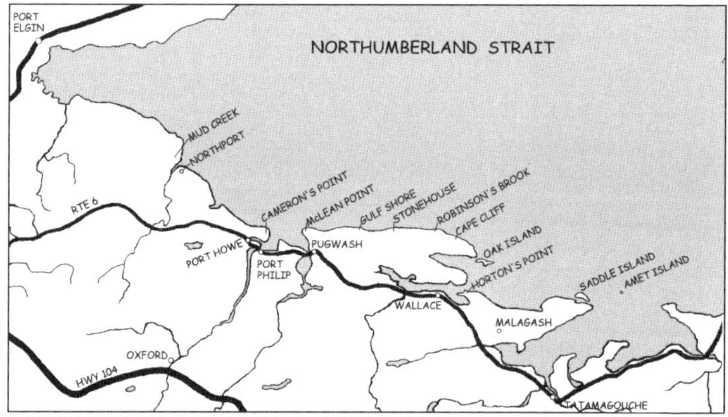

The North Shore of Cumberland County

Contents

Preface

The villages of Malagash, Wallace, Pugwash, Port Howe, and Northport have many stories to tell of the thousands of men and women who have gone out to the sea in small boats, often in hazardous and dangerous conditions, to trap lobsters, then to process and ship them to eastern ports in the United States and to cities across the sea.

These are stories that need to be told. How did our parents and their parents before them get into the fishing industry? Who built the many factories that dotted our shores and why were there so many? Why were their knockers as long as half a kilometre and why were their boats pointed at their stern? How did the tradition of Acadian girls working in factories arise and how did they get here from New Brunswick?

I grew up in Pugwash surrounded by the sounds and sights of the sea: everywhere lobster traps and boats hugging close to the wharfs, one trap above the other below, and the smell of salt and weathered rope that scents the air of a working gear. I gathered with other teenage boys and girls to swim in the harbour waters and along the sandy beaches that spot the shores of the Northumberland Strait and often in the sand we would find pipes or pieces of wooden structures that storms had moved and piled in their beating fury. Old timers would say these were the remnants of a lobster factory that had stood for decades, employing many through the hard times of the 1920s and 1930s, then into the 1950s.

I was delighted when the North Cumberland Historical Society approached me to research the history of lobster fishing from Malagash to Northport. I started almost immediately to interview fishermen — both men and women — who had been part of the industry for many years and who had insightful and colourful stories to tell. These are recollections of men and women about events that took place decades ago and, as would be expected, not every fisherman remembers the same events in the same way. Their stories are enmeshed with their own personal experiences and hence offer a unique perspective on the history of the fishery.

Although no one history can capture the enthusiasm and the personal style of the people who lived and worked at the heart of the lobster culture, I hope that some of the dedication and spirit of those I interviewed will come through on these pages. A sparkling dignity goes with the freedom and challenges of confronting the vicissitudes of a capricious and benign nature. The weather was always a concern: ice could appear early in the spring and damage trap lines, a nor'wester could smash boats against the shore for days, fog could descend and fishermen would lose their bearings. Lobsters might be plentiful one year and scarce the next, and prices could fluctuate with each catch. Fishermen struggled to find and keep the best lobstering grounds and were sometimes forced to violence when the fertile shoals were challenged by others. There were the hardships of combating the cold waters of early spring and late fall, the twelve-hour days of repetitive physical labour, the burden of hauling hundreds of sixty-pound traps, and the long weeks away from family. Yet, as one woman explained, "Once lobstering got in your blood it was there forever. The men talked and lived it. By February that's all the spring fishermen could talk about, even their dreams were about the big hauls they would make."

I owe a debt of gratitude to my daughter, Adrienne, who has been an indispensable guide to the use of language and the sequencing of a story. Any mistakes in language and spelling that might remain are my responsibility, for which I beg the reader's forgiveness.

Stephen Leahey

Malagash

Fred Brownell

Fred was born in 1915, the eldest of Ira Brownell's sons. Fred Brownell's mother had died, and he lived in Northport with his father, Ira, and siblings. He started fishing at fifteen to sixteen years of age with his father out of Northport. Ira had started fishing from a sailboat in about 1888 working both seasons summer and fall.

In about 1933 Ira and Fred put out their traps at Saddle Island at the first of the season. Saddle Island had the largest factory in the area and was started by Burnaham & Morril, from New England and taken over by Tuttle King. It was in the middle of the best fishing grounds and although there was no harbour on the island, boats found shelter from the waves on the south side.

Tuttle King leased the island and had about ten boats fishing its grounds from its shore. George Langille felt Tuttle had encroached on the fishing territory of his north and south factories, so his boats cut Tuttle's traps and began to destroy the buoys. In response, Tuttle King hired Ira, Frank Black and two fishermen from New Brunswick to go down and fish at Saddle Island and straighten things out. There were no laws at the time dictating where you had to fish. You could go anywhere in the district but convention held that you should limit your fishing grounds to the area of your factory, wharf, or packing outfit.

The first day at Saddle Island Fred and Ira put twenty to fifty traps on each of their lines and lay them along Waugh's

Shoals, which were noted for their good bottom. The line method had an anchor and buoys at both ends such that a boat could come in underneath the buoys and move along the line hauling traps. (Pugwash fishermen had moved away from the traditional line method adopting knockers, placing three to ten traps per knocker.) The next morning all of their lines were cut, although the traps were not destroyed. They gathered what they could. This destruction went on for three or four weeks as the Langille gang continued to cut the lines. One afternoon, around May 24, Ira went to Fred and said, "Put on your boots, we're going out." They cut all the Langille lines they could, letting the anchors and buoys go and, keeping so much of Langille's rope, the boat was loaded. Others were doing the same thing, including two New Brunswick boats.

The next morning Ira brought a whole handful of rocks on board and fished farther out than usual. At the end of the day Ira and Fred ventured back to Waugh's Shoal and saw that all of their buoys were gone. They threw grapples over the side of the boat to try to get as many traps as they could. At the same time five Langille boats came alongside, two to each side and one across the bow. The people in these boats made a lot of threats: one guy said he was going to get his rifle and blow the head off anyone who came back to cut buoys.

This to-and-fro went on for three or four years. They would put out 500 traps and land only 300 to 400 at the end of the season. This was very expensive, but Tuttle King guaranteed his fishermen a certain dollar independent of the size of their catch, and this helped to offset the loss of their traps.

One day when Fred was about twenty years old, his aunt, Mrs. Hilda Trenholm, came to the door wanting Fred to work with her son Hudson trucking lobsters from New Brunswick. Fred accepted the job and stayed the winter with the Trenholms helping Alf tend his foxes. Fred fished the spring of 1936, and then worked at the salt mine for the rest of the year until the following spring. Throughout this period, Fred lived with the

Trenholm family and did odd jobs for Alf. By that time Alf had aged and left the lobster business.

One time, in 1935, Tuttle's men looked out and saw a guy who looked like he was cutting their traps. Ira, Fred and the two Acadian guys went out to confront the man, but it was a guy who turned out to be a friendly person. Right then a boat came around Amet Island shooting at them. As the shooter came closer, he fired two more shots to scare them. The Acadian guys fell to the floor "counting their beads." Later Ira took his rifle down to the shore and fired a warning shot at a boat that seemed to be fishing his traps. The boat came into the wharf and one of its men went into the factory, but Mrs. King, who was a large woman, grabbed him by the throat and pushed him back to the wharf. The next day the Oxford Journal ran "War Declared on Saddle Island."

Whenever Ira fired his gun he would report what he had done to the RCMP. The constable would talk to Ira in private, but never did anything and always left smiling. There was some suspicion that Ira, like other fishermen, was running rum — meeting boats offshore, loading up with their rum barrels and bringing them to shore — and that the constable had caught but not reported him. As a result, Ira had some leverage in his negotiations with the constable.

Fred remembered the rum running in Northport when he was young. One of the main runners was an Alie Philmore from Port Elgin. The rum was brought in by night and came from St. Pierre. One time Ira went to St. Pierre on his forty-foot fishing boat, but more often he was just one of four or five boats that would bring the rum barrels in from offshore boats. Frank Black also helped bring in the rum. Each boat was paid $100 a night — a substantial amount. On one occasion the fishermen landed rum at Mud Creek at night and took the rails off the bridge to offload. Fred knew the fishermen had been out getting rum by the barrel rust rings on the floorboards of the boats.

By this time, Fred had his own boat, purchased when he

was twenty-one years old. He bought a used boat and it didn't have a hauler; instead his helper, Lorne Black, hauled the traps hand over hand. Fred had a hauler on order at the start of the season and it was delivered one day as they were fishing. He could see it on the shore of Saddle Island. Lorne was ecstatic at seeing it resting idly on the bank. The bright red make and break engine made a bang and was attached to a cog connected to a hauler. The hauler was standard equipment until the 1980s. It wasn't difficult to get a license at the time Fred got his first boat and there was no limit on the number of traps that you could use — 500 to 700 was standard. The price of lobsters was seven to ten cents a pound. One season the price was so bad that Fred saved up his lobsters, 600 to 700 pounds and took them to the United States in a half-ton truck. In 1936, Fred moved from Northport to Pugwash, living there throughout the year. His aunt, Alf Trenholm's wife Hilda, asked him to be her son's helper in the fall season.

The fighting between Langille and King continued. Sonny John McKay, a resident of Saddle Island, used to watch the Langille gang fairly closely. One night while they were cutting the Saddle Island fishermen's traps, Sonny went down with his old 57 Synder rifle and shot at them from below the factory. The bullet didn't reach its target but the bullet's rust left a stream over the water and caused much laughing. Fred and Lorne some-times followed Langille's boats to ensure they were not cutting lines. Lorne was good at throwing fish, one time he caught one of the "enemy" right in the mouth. It got Langille's men "mad-der than a hatter." On another occasion a guy by the name of Harold Walton was out on the water, following the Langille boats around. Unbeknownst to Harold, one of these boats had a Mountie onboard. As Walton approached the other boats he snapped his engine off and showed his gun laying on the engine house. Up came the Mountie and said, "Pass it over " and seized the rifle. Eventually Harold got it back because he said he used it to hunt seals.

By 1939 Clarence Kennedy had taken over from the Kings on Saddle Island and Fred was running the smack boat for Kennedy's lobster factory in New Brunswick. It was also the year that Saddle Island burned to the ground — nine of the good buildings, traps and boats with them were gone.

Kennedy decided he had to do something. He hired an ex-RCMP guy, Arthur King, bought him a brand-new car and set him out to hunt down the arsonists. King worked at it for three months finally catching a Langille fisherman. The case went to a trial that lasted four to six weeks, and the guy was convicted and Langille was charged. Langille was cleared eventually but it cost him and his sisters a lot of money and it was rumored he was cleared because he was a Mason.

There were still some fishermen and shacks left on the island and a new cook house was built, but the fishing battles were over. That winter Fred and his sisters knitted all the heads and rebuilt all the traps that had been lost, but moved his boat to the Malagash wharf and fished from there.

Typically, in the spring season, while working out of Malagash, Fred would leave Pugwash on Sunday night, work in his shack, and come back on Saturday afternoon. The fishermen's shacks were eight by ten feet and had upper and lower bunks and a wood stove. People would eat out of the cookhouse. Mrs. Reeves was the last cook Fred remembers at Malagash. While most shacks had passed from fisherman to fisherman over the years, Fred built his own. Not everyone had their own shack, however, and some would sleep in the bunkhouse. Fred hitched a ride to and from Pugwash until 1946 when he bought a 1938 car. In going and coming on the Stake Road he often became stuck in the ruts. These were very bad and the road remained such until Langille, a good Tory, had all the roads in Cumberland County paved in about 1973.

Fred fished the spring of 1942, and then hauled the boat up near where the Pugwash Legion is located today, and went to war for four years. Alf Trenholm's factory on the Gulf Shore used it for the following spring seasons, but then it was smashed on the rocks at Robinson's Brook.

There used to be many lobster factories: one where the Pugwash Yacht Club is today, one where the snowplow shed is on the road to the Gulf Shore, one at Pugwash Point and another opposite the Northumberland Links golf course, where the Hollises live now. Many of the gears in the pre-1941 days were owned by the factory owners. There were not enough people locally to handle the workload in the factories. Most of those who worked in the factories came from the South Shore and East Shore (Guysborough County). Some even came from New Brunswick, mostly the girls who packed the lobster. The girls came from the Bouctouche, Cape Ball and other Acadian areas and were often chased by local boys. Boiling the lobsters and certain other jobs were done by local men.

Fred was out of the Army in February 1946 and fished that spring season. It was very challenging to refit all his gear: his trap heads had to be replaced after three or four years and his funnel heads were worn out. Fred remembers his wife, Bessie, spending many nights knitting heads along with him. His brother Cecil fished with him and together they had about 400 traps. Alf Trenholm gave them a boat and then they bought a thirty-five-foot pinkie built by Main Allen in Port Elgin for $125. In about 1948, Kennedy moved many buildings over the ice from Saddle Island to the Malagash wharf.

Fred fished in Malagash until 1981, a total of thirty-five years. Fred remembers the best Malagash fishing spots: Waugh's Shoals, Abbott's Sound, near Abbott Island, and Trenholm's Shoal, about six miles offshore between Malagash and Wallace. However, the fishing grounds at Malagash are no better than those at Pugwash. The success of a good fisherman depends upon his skill in finding a good bottom, fishing more days than others and taking good care of his traps. Before the advance of electronics a fisherman had to grapple on the bottom to see whether they had sand, seaweed, or rocks. Lobsters prefer rough bottom. Not even the Department of Fisheries knows where the lobsters come from. You can even catch them a hundred miles offshore but the lobster shells are very thick because of the water pressure. Fred moved to using finders in the late 1960s. Long before then there was an outfit from Liverpool that had a large clumsy depth finder that used to send clicks but it didn't tell you the type of bottom, just the depth.

Now the great majority of the lobster factories have been closed and the lobsters go to a bigger centre for processing, although it appears there is still some processing done in Wallace. The price moved up and down and was pretty poor most of the time. Most fishermen couldn't make a living by just fishing lobsters so they went to the lumber camps in winter. Fred used to pick up the odd job but never worked in the woods. He worked in the Malagash mine one year and fished smelts down Wallace Bay and near Victoria Island.

In 1981, Clifford Allen convinced Fred to move to Pugwash, as it was more convenient, rather than traveling back and forth to Malagash. A new fisherman in Pugwash needed the support of eighty-five percent of the other fishermen; Fred passed easily because of friendships that he established over the years. Today the Pugwash fishermen are not allowing anyone to come in, and now when you have a license in Pugwash that is where you have to stay. There are now about fifty percent fewer fishermen. The government wharf in Pugwash used to be lined with lobster traps, but this is not the case now. Today there are two types of lobster licenses: a full A license in the spring entitles you to 300 traps and in the fall to 250 traps; a B license entitles you to 90 traps in each season. A full gear today would be very costly: 300 at $80 apiece would total $24,000; a fiberglass boat would be at least $150,000, instruments would cost another $15,000 and the license would be over and above that.

In 1985 the fishermen began to use wire traps. The new traps caught more lobster. Fred started with 200 wire traps and caught thirty percent more than he had with 200 wooden traps.

In 1986 he built a full gear of wire traps. Although a fisherman is supposed to have his own traps, it was not uncommon for the men to share traps between the spring and the fall seasons. Wire traps brought in the biggest loads Fred ever had. His best year was 1988 when he landed 45,000 pounds of lobsters.

In the 1980s the Department of Fisheries stopped fishermen's practice of taking gear out in the beginning of the season on two boats. There were often clashes between the fishermen and the Department of Fisheries. One time a fisherman got intoxicated and punched out a window in the DFO cutter. He would have bled to death had someone not been there. He was eventually cleared by Tommy Giles, a well-known lawyer.

Fred lives in the same house he bought when he married Bessie. He left the fishery in 1991 and passed his license boat and gear to his son-in-law Stephen Ferdinand, Howard Ferdinand's son, thus perpetuating the fishing tradition of two distinguished families.

Charles Farrow

Charles was born on December, 28, 1918, at Cameron's Point in Port Howe. Charles remembers taking horse-drawn dump carts to meet the lobster boats and off-loading the catch. The sailing lobster boats didn't have to go too far out to catch a thousand pounds on which they were paid a dollar per hundred lobsters.

Charles's father went to work at the Malagash salt mine in 1923 but didn't move his family there until 1927. That same year the Malagash salt mine railroad was built from Malagash Station to the mine. It connected with the C.N.R. Shortline, which extended from Pictou to Oxford Junction, with stations at River John, Scotsburn, Tatamagouche, Wallace, Pugwash Junction, Pugwash, and Oxford Junction. The Shortline Railway was built in the late 1800s.

The shoreline between East Wallace and Malagash was dotted with small lobster factories. Some were located by sandy beaches that allowed men to pull their small fishing boats ashore. These beaches were also popular spots for group and church annual picnics. Where factories had to be located on a rocky shore, pole wharves had to be built each spring. Every factory needed a source of fresh water, whether from drilled wells or brooks, for the steam boilers and to cleanse the lobsters. Days at the factory were very long, sometimes around the clock, and workers would almost fall asleep on the job.

George Langille had two factories: one on the north shore

and another on the south shore of Malagash Point. A prominent packer by the name of Tuttle King operated a very large lobster station, with many boats, during the 1930s. His fishermen were mostly from Cape Tormentine and Port Elgin in New Brunswick, Northport, and Pugwash. Betts' factory was on the south shore at Treen Point. A lobster war broke out on Saddle Island between Langille's people and King's on the island. Both groups fished long lines of traps and regularly crossed and cut one another's lines, resulting in large losses of traps. In 1938, Saddle Island was set on fire and burned to the ground. Charlie Corkum from the Eastern Shore was convicted and sentenced for two years but then released early for good behaviour and for volunteering to join the war. He was paid to set the fire, the money having passed through several hands so that the main culprit was never caught. A second suspect was tried and found innocent.

Burnham and Morrell were the local packers and they shipped most of their product to Halifax.

The Kennedys took over Saddle Island then moved its buildings over to the Malagash wharf when the mine closed. The Kennedys used the large abandoned mine warehouse on the wharf to store their boats and gear. It later burned down. Clarence Kennedy was a prominent Malagash packer who had originally come from Rockley, near Port Howe. His wife ran the cookhouse for his factory. The Kennedys refused to go to daylight saving time in their factories for religious reasons.

Clarence, had two sons, Ken and Everett, who was the black sheep of the family. After the war Everett bought a Fairmiles motor torpedo boat (M. T. B.) and used it to bring bait and lobsters from Newfoundland. The family was quite religious but Everett liked to drink and act up a bit and was noted for being a very fast driver.

In later years, permanent homes and summer cottages were established on the old factory sites: for example, the Blue Sea Park at Treen Point, Malagash, occupies the land where the Betts' lobster factory once stood.

There was always a bit of excitement when families came from the Eastern Shore to fish or work in the factories. They journeyed from Sheet Harbour, Ecum Secum and other area villages by train to Oxford Junction, from which they would take the Shortline to Malagash Station, where they would be met by horse and wagons. They worked mostly in the two Langille factories or in the Betts factory, and many stayed behind at the end of the season and married into the local community.

In the 1920s, with the advent of motorcars and railways connected to New Brunswick, the South Shore women were replaced by Acadian girls from Cape Ball, Shediac and Bouctouche. The girls were Catholic and, since there were no Catholic churches in Malagash, there was no intermarriage. The girls were single but had a matron to chaperone them. Because there were as many as a hundred single men working at the mine, the arrival of these women caused much excitement. After work, the guys would take soft and hard drinks plus chocolate with them to meet the women. Some girls did not speak English but were able to communicate by sign language. The men would walk or travel by horse and wagon over the muddy two-and-a-half mile road from the mine to visit the girls at the lobster camp and hold dances in the cookhouses.

After the fall season, several fishermen in Wallace and Malagash would fish at night and take the lobsters back in the woods to cook and can them using labels from legitimate operators. These operators would pay the fishermen and their families for these cans and place them with their spring lot. Alternately, some packers kept cans of lobsters and tomalè in their basements and sold them as demand warranted. Fishing inspectors turned a blind eye to these activities as it was considered a common way of life. Illegal fishing continued for many years with lobstermen selling fish in the hard shell if they couldn't can them.

John David Reid was a fishing inspector in the 1920s. He had lost a leg in the lumber mills and traveled in a wagon led by a white horse. The illegal fishermen could see him coming from a long way off and often took evasive action. Man Trenholm, from East Wallace, was fishing illegally — by federal standards, not local — and had a few mackerel nets out. Seeing John David Reid come into his yard, he hurried ashore and after an interminable length of time greeted John and invited him to have a drink. John, who had a fondness for rum, readily agreed. After a couple of drinks Man pushed the bottle towards John and suggested that John make himself at home while he went out to his nets for some fresh fish. John agreed and Man went out to his lobster pots, always keeping an eye on John and the white horse. After a couple of hours Man returned from fishing all of his pots and nets and found John sound asleep and snoring beside an empty rum bottle. Man got John David into the wagon, untied the horse, slipped a couple of mackerel beside him and pointed the horse down the road and to home, and turned to boiling his lobsters.

Cans were soldered with lead. The soldering iron was shaped to fit the curving of the can. Canning machines did away with soldering in the early 1900s. A sample soldering machine is on display in the Malagash museum. In the absence of electricity canning was done in the factories using a pulley arrangement driven by steam. Electrical power came to Wallace and Malagash in 1930. The power station on Waugh's River supplied the electricity and sales were made one at a time.

Before modern canning and cold packing came along, lobster families often bottled lobsters in brine for home use. Charles remembers one older guy reflecting on hard times in his childhood saying: "Sometimes things were so difficult all we had were lobsters."

Some lobstermen had farms but they were not particularly good at farming. Their main skill was fishing.

Rum running was a related sideline in the prohibition period in the 1920s and early 1930s. A sailing vessel would go to the West Indies and trade salt fish for sugar, molasses, and plenty of rum. The ships would bring loads of liquor in various size barrels and kegs and lay off the twelve-mile limit in the Northumberland Strait. At dusk, they would move into channels close to land and local fishermen would sail or motor out to meet them and ferry the cargo to shore. Sometimes captains from Port Elgin, New Brunswick, as well as from villages and towns along Nova Scotia's Northumberland shore would sail as far away as the French islands of St. Pierre and Miquelon, off Newfoundland to pick up liquor. Once ashore, cargo containers were hauled further inland by horse and wagon, then transferred to cars and trucks. Rum and molasses would be taken from these ninety-gallon drums and split among many five- and one-gallon jugs. The molasses would find its way into stores and the rum would be distributed to bootleggers.

In 1927, a rum running schooner came directly into the Malagash wharf. Two men aboard spent the afternoon wrenching the ninety-gallon puncheons onto the pier and that evening a convoy of large half-ton trucks came and took them away. At that time, large trucks might be not much larger than a half ton. It was said that the trucks were owned by the Chicago syndicate run by Al Capone. The police at that time were few and far between.

Burt Langille

Burt was born on January 30, 1927, in Malagash and went to the local school, just two hundred yards up the road from his family home. His brother Winston was born in 1940. When Burt and Winston were growing up there were sixty-three people living on Malagash Point. Now all of the original inhabitants are gone. In the 1930s and 1940s people did not travel very far. Burt was fifteen years of age before he got as far as Tatamagouche. Winston recalls that when he went to the regional school in Tatamagouche there were only five students at the Malagash School. He was highly intimidated by the relative crowd in his new school.

Burt remembers making hay with Alex McKay. Alex was paid a dollar a day but Burt, who worked just as hard as Alex, got only fifty cents as he was still younger than twenty-one. People used to go to Oak Island to make hay and while there they would fish and can lobster illegally. On one occasion, an intoxicated inspector went to the island at low tide by horse and wagon. He looked around, fell asleep on the couch, and left. He had completely missed the snapping cans as they cooled off under the steps of the building. There was a lot of illegal fishing in the thirties because times were hard and people had to get enough to eat. They would fish at night. There weren't many fishing inspectors and even if the fishermen were caught the fine was very small. Some people used to land rum on McLean's beach. In the early 1930s, many of the fishermen would carry guns on board their boats; People from Prince Edward Island would sometimes try to take. over local lobster fishing grounds.

Burt saw the German Zepplin going up the Northumberland Strait in 1937.

In the late 1800s, Lloyd Boyd Whitting established the first factory on Saddle Island. By the turn of the century there were fifty to sixty Acadian girls working on the island, and a lot of Acadian men came as well. They would have dances every Saturday night. On September 15, 1938, Burt watched as flames engulfed Saddle Island. Tuttle King from Tatamagouche owned the factory at the time. Clarence Kennedy took it over after the devastation. In 1948, the ocean around the island froze so solidly that Clarence Kennedy was able to tow his factory buildings over the ice to the Malagash wharf. It was the only time in memory that it was cold enough for the passage to freeze completely.

A ship bound for Newfoundland and loaded with cattle ran up high and dry at Langille's factory in 1952. The factory girls came out in their sleeping attire to watch it wait out the tide.

Burt used to work at George Langille's factories for a few days at the end of each school year, after the Acadian women went home and the height of the lobster season had passed. In the 1940s the girls would get seven dollars a week plus room and board.

George Langille had two factories: one just up the road from where Burt now lives and the other on the south shore of Malagash Point. Ted McCollum had a factory at the end of Malagash Point and Alfred Betts had one where the Blue Sea Park is today. Gordon McInnis had a factory on the Gulf Shore at one time. The Burt Langille fishing grounds were named after a man who fished the area but not related to Burt. The Betts grounds were named for Betts, whose factory was nearby. There were lots of lobsters to be canned and factories flourished. Lobsters were so inexpensive that he remembers paying George seven cents for a one pound can of lobster paste.

A big yellow Anson bomber crashed and plowed up into the beach in 1943. Burt was picking potatoes at the time and saw it come down. The pilot and his passengers were safe and went to the Saturday night dance. The plane was eventually towed over to Saddle Island and left there.

The "nigger head," used for hauling traps, was really a "knotter head." A snood knot, which tied the line to the trap, was something like a half-hitch knot. When this knot hit the hauler you had to flip the rope over the hauler quickly or else the trap would knock the boat. When the fishermen fished with lines, they had to stay in the same spot to leave lots of sea bottom free for the lobsters to breed. Fred Brownell was one of the best fishermen Burt ever saw. He always got the most lobsters and kept a very tidy boat.

Burt fished the Malagash north shore until 1962 at the

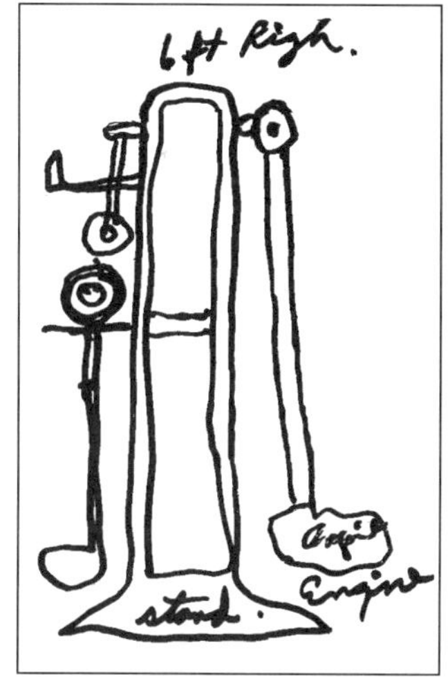

Above is a hand-drawn picture by Burt Langille of a canning machine used in George Langille's factory in the thirties. At the time health rules forbid soldering so the lids had to be pressured on to be airtight.

The machine weighted about a 175 pounds and took two men to lift. The circle part could handle three sizes of cans. An engine spun the top. The lever on the left was operated by foot. The cans were placed on the circled part and the cover pushed on. The machine handled fifty cans a minute.

same time as he worked full-time at the salt mine. He had bought a license for twenty-five cents and had a fifteen-foot boat and about a hundred traps. Lobsters were selling at the time for twenty-eight cents for canners and thrity-five cents for markets. He retired from the Salt Mine in 1969 to raise foxes at his farm in Malagash.

Douglas Mills

Doug was born in Rockley in 1913. He started fishing smelts in River Philip behind his house when he was fifteen. A few years later Lloyd Simpson, a fellow smelt fisherman, encouraged Doug to fish lobsters with him from Josh Allen's factory at Horton's Point in East Wallace.

When Josh sold the factory to his brother George two years later, Doug moved to the Kennedy outfit on Saddle Island. Doug knew Clarence Kennedy, who was originally from Port Philip, many years before. The Kennedys owned Doug's gear and boats and his first helper at Saddle Island was Lloyd Reid.

Saddle Island was quite small and when the tide was out you could almost walk ashore. Doug stayed on the Island for only one year, then moved to the Malagash wharf (although they loaded gear from Saddle Island the second year). A bunk and cookhouse remained on the island even when they stopped canning there.

The Kennedys owned all the lobster gear and boats. At the end of the season, the boats were taken to Bayhead and placed on the Kennedy property, while the traps and related gear were stored in a building on the island. The men usually arrived at the factory on April 15 each year, ready to make any necessary repairs to the traps. One day was set aside for turning the boats right side up. They had been turned upside down after a drying period the previous fall.

Clarence Kennedy had all his houses and warehouses on

his property at the wharf and even built a new office there. For many years he remained a stickler for Standard Time. As a result, the men were required to have breakfast at 4 a.m. Eventually, Clarence told the men that if they wanted to observe Daylight Saving time they would have to breakfast at 5 a.m. This breakfast hour remained in effect throughout Doug's time at the Malagash wharf.

Boiled eggs, pancakes, toast and prunes were typical breakfast fare. The lunch contained sandwiches made with homemade bread. Bologna sandwiches were a favourite. Cookies were always added as a treat. Mrs. Reeves was the cook and was from Malagash. Meat and potatoes and lots of vegetables with cake, cookies, pies and rolls for dessert composed a usual dinner served from noon to 4 p.m. depending on when the men returned from their day of fishing.

When Kennedy's men moved from the island, the Malagash wharf went straight out into the bay. The Malagash Salt Company owned a large storage building on the wharf, from which salt was loaded on small ships for transport. Clarence Kennedy bought the building as well as adjoining lands upon which he located his cookhouse, bunkhouse, office and additional storage buildings. The original salt building stored all the traps. The building later burned down with all the traps in it. The cause was a mystery, but suspicion fell on arson. In time, the Kennedys and local fishermen got the government to build an "L" part on the end of the wharf, making it possible for fishing boats to shelter from storms coming off the bay.

One afternoon, as the fishermen came ashore, the wind was coming down the bay very hard. They anchored but Clarence asked them to bring his boats up near the shore. Some people didn't do this and their boats were badly damaged. Doug's boat, along with two or three others up near the road, survived unharmed. On another occasion, Doug and his helper were fishing off Amet Island in May. They were hauling in their knockers and so had turned off the main motor. When Doug went to start the main motor again it wouldn't kick in. The wind was blowing so hard and the water was so deep that the anchor couldn't reach bottom. Thinking quickly, Doug pulled the anchor up and used a spare coil of rope to extend the reach of the anchor and threw the knocker overboard to further secure the boat. Despite these efforts, the boat continued to drift, buffeted by waves until sunset when the wind subsided. The temperatures were low and they spent a very cold night. Early the next morning Clarence and several fishermen arrived with lots of hot tea and food and a mechanic to repair the motor.

The fishermen got along well, but once when Doug fished a little too close to Amet Island, someone would pull his traps and mess them up. Another time, the ice was late breaking up and it was agreed that fishing would start on a Monday morning. Nonetheless one fisherman took two loads of traps out on the Sunday. In retaliation, someone took a crowbar and punched a hole in the man's boat causing it to fill with water. The fisherman managed to pull his boat ashore and got the hole plugged.

Doug enjoyed fishing and cannot remember disliking any part of it. He continued fishing at Malagash until 1969, when the lobsters became very scarce.

A neighbour up the road had a Gaspereau operation on River Philip. Doug and Jean decided to buy it, selling the lobster gear and entering the gaspereau business. The first year they packed 900 barrels with the help of four men. The gaspereau was salted, closely inspected by Nova Scotian authorities, and shipped to Haiti. Doug sold the operation in 1993, when he was eighty years old.

Howard Treen

Howard was born on December 15, 1927, in Malagash, just below where the Malagash wharf currently stands. Howard went to the nearby Malagash Centre School but left in Grade Nine, at the age of fifteen, to work in the Malagash salt mine's warehouse and wharf. His father was a farmer who also fished herring in the Malagash Bay and had a smokehouse from which he sold his garden produce locally. Howard began to fish in the Bay while working at the mine. With a small outboard motorboat and a hundred traps, he was able to bring in about a hundred pounds of lobsters per day.

In 1947, Howard started fishing on a regular basis as a helper to Mel Langille, fishing from the Malagash wharf and selling the catch to Clarence Kennedy. Their boat was twenty-four feet long and anchored in front of Mel's house, near the wharf. The mine still used the wharf full time to receive salt by rail from Malagash Station. Tying boats up at the wharf posed some unique challenges as it was built so that the tide and waves could move water freely between one side and the other. The "L" part of the wharf had not been added and so it had no sheltering alcove. Most independent fishermen landed their catches on the wharf but anchored off the shore. George Langille's north shore factory was still processing lobster at that time,, and he, too, had boats anchored off the south shore. The wharf was eventually "rocked" and then later transferred from the salt mine authorities to the local fishermen.

In May, 1950, a large storm flooded and dragged eight to ten boats ashore even though they had been well anchored. Russell Elliott's boat was the only one that didn't drag ashore. The others came in on the sand and were salvaged intact, including Howard's. By then the fishermen owned their own boats as there was a movement at that time for the fishermen to become more independent.

In the mid-1950s Howard worked with his brother Harold in a Kennedy boat. Clarence Kennedy, the buyer, was a great guy to fish for. If your motor needed a part he would quickly go off to Halifax to get it. No one had a bad word against Clarence. Except for a couple of years when Charles Kennedy sold his father's business, the Kennedys have been permanent fixtures in Malagash.

The River John Co-op began to compete to buy lobsters in Malagash in 2003.

Clarence was an old sea captain who lived at Bayhead. One year there was a week of fog, and the fishermen left in the fog and came back in the fog. They would find their knockers by compass and move from one knocker to the next, following a close line as they were all placed close to one another. At times, the fog was so thick it was in the boat. Clarence Kennedy, said he didn't now how the fishermen did it.

The Kennedys moved their factory, bunkhouse and cookhouse from Saddle Island to the Malagash side. Howard stayed there for a few years prior to getting married so as to save his mother the time and expense of housing him at home. The fishermen at the wharf got along well among themselves. There was a time, however, on Saddle Island when fishermen carried guns aboard their boats.

Howard and Harold fished together for several years and were eventually able to purchase their own boats. In 1974, Harold was killed in a head-on collision driving back from New Brunswick where he had bought some herring nets. During the really poor seasons, in the early 1970s, Howard fished without a helper because the season's take would only support the wages of one man. He remembers an opening season when Fred Brownell caught only seven lobsters and asked, "Do you want to buy a gear?"

Howard really enjoyed fishing; it got in his blood. To be on the water was a thrill. Even after retiring in 1995, he still dreams of fishing.

Wallace

Harold Elliott

Harold was born in Port Howe in 1929, and comes from a long line of lobster fishermen, including the Camerons, who had a factory down Green Lane in Port Howe. The area upon which the factory sat has been claimed by the sea, with nothing showing but pipe sticking out of the sand there now. Harold remembers his father saying that, in the early 1900s, they would look in the gullies along the shore for lobsters. When they found all they needed for dinner under the seaweed and among the rocks, they would fill a bushel bag and return home.

Harold helped his father at McInnis's factory from 1947 to 1951. His associates at the factory included Charlie Rafuse, Earl Pye, Thorpe Moody, Harold Jamieson, Roger Roblie, Murray Smith, Lester Myers and Lester's father. Harold believes that McInnis obtained the fifty-year-old factory from either his father or his uncle. It included a large bunkhouse that was divided between a place for the men and a place for the twenty or so women brought in from New Brunswick to process and can lobster.

The packers took advantage of natural shelters when deciding where to establish a factory. The factories were generally located along the shore in the lee of an island or natural reef. These areas were sheltered from storms and served as safe mooring or anchoring areas at a time when there were no wharfs big enough to accommodate the fishing boats. The boats were too small and lacked the power to make the long trips to harbours such as Pugwash and Wallace, and poor spring roads kept the men stationed at the factory, prevented from traveling back and forth to their homes. The Oak Island factory was located at the end of a so-called duck pond, which protected boats from northeastern storms. Everyone fished out of predominately company owned boats and stayed at the factories for two or three weeks at a time. Some who came up from the Eastern Shore would stay the whole season.

The Acadian girls who came down were poor, for the most part, and looked forward to the excitement of the season. Surprisingly, even in the early 1950s, the factory lacked running water and both the men and women had to wash from barrels of water. In many respects, the factory had not changed substantially since its inception at the turn of the century. It lacked electrical power, but had a telephone.

The factories were small and could only handle a limited amount of processing. On days with very large hauls, even with everyone working full out, there would be considerable wastage, as they had no means of keeping the lobsters fresh. This added to the already high wastage caused by only using the tails and claws of the lobsters. The cannery evolved and changed this so that by the early 1950s meat from the legs and the knuckles, in addition to that from the tails and claws were extracted and preserved. Moreover, the lobster liver, or tomalè was used and sold separately. Tomalè used to have a lot of meat in it. Today it is thin and goes by the name of lobster paste.

The boats were small in the early days of the factories and relied on a sail and oars. These twenty-foot sailboats had a centreboard that could be pulled up as the boat came to shore making it like a dory. This was particularly needed in times of storm. Many of these boats were made in "lap work," or overlapping boards, which served to steady them in rougher waters. The boats seldom went more than a mile or so off shore. It often took the sailboats days to get the traps out, even though they didn't go far from the factory. The trap lines could have as many as a hundred traps on them and run for maybe a third of a mile or more. It took successive trips by boats loaded with traps to build one trap line.

Pinky and V sterns were popular in that they could move forwards and backwards along the trap line. The lightweight traps shifted with the waves so that their shoes would be worn thin at the end of a season or two. Knockers with maybe twelve traps per line were adopted in the late 1920s, and eventually boats used car motors for power. The first motors were single cylinder and would go only about six knots per hour. Modern boats will go three to four times as fast. Now with the electronic systems that are on every boat, the lobsters don't stand a chance, much different from the old days when it was more of a hit-or-miss business, especially for new fishermen. Polypropylene rope and nylon twine has also been very cost effective for the fishermen because they do not rot or need replacement nearly as often as the conventional manila rope and traps, which had to be thoroughly dried before they were put away for the next season. The cost savings realized allowed some of the fishermen to stay in business during the hard times when catches were low.

In about 1950, McInnis closed the factory, keeping the fishing operation, and sold lobsters directly to Josh Allen. Gordon Darragh took over the stand and lobsters were trucked out daily over a road system that threw up many hazards and delays, particularly in spring. As the number of factories declined, some of the fishermen sold to the smack boats. These boats would meet the fishermen on the strait during the day or in the evening after their catch was in, load up and take the lobsters directly to Josh. In some small measure this was an attempt to bypass the terrible road conditions during the spring season. The last factories into the fifties were George Allen's in East Wallace and Kennedy's in Malagash.

During the forties and fifties a lot of the men fished both spring and fall seasons and paid the fine for doing so. Sometimes the fall fishermen ran a long line into the spring territory, grabbing the lobsters that had been too small or to soft to take in the spring but that now after a few months were primed for fishing. The spring fishermen were unable to capitalize on a similar situation. Company packers held the fishermen through credit, often extending it through the winter in the years before Unemployment Insurance. In many instances the packers also ran winter lumbering camps so that they might employ the same men in both places.

In 1952, Harold worked with his own helper for Josh Allen out of Pugwash. Harold fished every spring season until 2001. When Paturel bought out Josh in 1959, it sold a number of his gears and Harold was quick to buy one. Harold bought a new boat in 1965, calling it *Kitty E.* In 1978 he purchased another boat, *The Souwester,* which was wooden, had a diesel engine and was forty-foot long. He purchased his next boat in 1989, and named it *Steve B.* It was thirty-six-foot long but wider than the previous boat, and was made of fibre. On April 15, 1999, Harold lost *Steve B.* and 230 traps to a fire that also claimed another man's boat and 490 traps. The fall fishermen came to his rescue with a boat and gear and asking little or nothing in return.

Everything changed in the 60s, from factories to shipping lobsters live to their American or European destinations. Harold's observations and insights include:

Harold has built wire traps since 1985 and holds several patents on traps and accessories. His brother Alvin had demonstrated their worth to him in 1984. The next year they both went down to Maine with two trucks and loaded up with wire. There was an immediate demand for the product: Brud King went from all wooden to all wire in one year followed closely by Gary Bollong and Ronnie White. The lobsters were good then and the wire traps went over well. Almost everyone uses wire today, although the lobstermen on the North side of P.E.I. have held strongly to the wooden variety.

In 1988 Harold incorporated his business. He has had as

many as six workers building traps. His peak sale years would have seen the sale of 6000–7000 lobster traps in addition to his crab and eel trap sales. A good wire trap will last you twenty seasons whereas a wooden one will give you only about fifteen, during which you'll likely have to replace about fifty percent of the trap along the way. If you have two traps, one wire and the other a well soaked wooden one, the wire one will weigh thirty-five pounds and the wooden sixty-nine pounds. Under water the wire one weighs about twenty pounds and the wooden one about nine. Being lighter out of water and heavier in the water gives the wire traps some major advantages, including in rough weather a buoy will pull the wire trap less and better allowing a nervous lobster to enter.

Harold believes there is a cycle to the supply of lobsters. Over-fishing in the late 1800s led to a scarcity of lobsters, and happened again in the 1920s and later in the 1930s. The 1940s and 1950s were good years for the most part with the lobsters peaking in 1959. In the early 1970s, the catches reached a very deep low. Harold has seen two peaks in his fifty-five years of fishing; the one in the late 1950s to early 1960s and the other in the mid 1980s. During the poor seasons, Harold would fish alone, and some days either not fish at all or go and work in the woods. This cycle has led him to believe that a young man starting out will see two peaks and lows in his lifetime and that the best time to enter the business is in the low part of the cycle when boats and gear can be purchased cheaply. When the lobsters are at their best you renew your boat and gear letting them slide when the lobsters taper off.

The fishermen like the suspense of waiting and seeing what will be in each trap, especially when the fishing is good, and they enjoy the competition of getting to a rock reef first and trapping the biggest and best lobsters.

After fifty years as the captain of his own boat Harold still has a vivid memory and a map in his mind of all the places he fished. His son fishes now out of Wallace.

Lloyd Jameison

Lloyd was born in Wallace in 1918 and in his teens worked as a carpenter and on his father's farm.

In 1938 he fished with Harold Jamieson out of George Langille's south shore factory in Malagash. George had roughly six boats out of both his north and south shore factories and employed about twelve Acadian girls and a cook. One cook who was widely praised was Hardy Coolan from the Hubbards area. After the Second World War, he cooked for Gordon McInnis for a couple of years.

Lloyd remembers a story about the factory at Horton's Point. John Charman, (Millard Charman's father), Mann Trenholm, Josh Allen, George Allen and the Conlew Company all owned the factory at one time or another. Two young men, Cyril McInnis and a Vincent boy, were wrestling and carrying on. The floor was slippery and they fell into the boiling vat used to cook lobsters. Vincent was on the bottom and died. Cyril, Merril Jamieson's great uncle, was burnt badly but lived and eventually moved out West.

Lloyd fished for a season from Saddle Island in the early 40s. He worked on Russell Elliott's boat, which was owned by Clarence Kennedy. Shortly thereafter Lloyd went overseas in the War. When he returned in 1950 he fished again for George Langille out of the south factory. The size of the factory was still about the same as it was in 1938. Lobsters were plentiful enough to keep the factory girls busy.

There were no factories in Wallace in older times. Shore factories, such as those in Malagash and on Saddle Island, were established because the boat engines could not power a craft from Wallace out to the lobster grounds and back fast enough to timely cover the five miles from Wallace to Oak Island. Isaiah Whitman, an older fisherman from the Gulf Shore, quit fishing when the motor boat was introduced because he didn't want to give up his sailboat. He was a careful craftsman, and in later years Gordon McInnis had Isaiah make all of his traps. In about 1960, when motors became more powerful, boats began to fish from the Wallace wharf. The Kennedys, the Paturel Company, E. P. Melanson, and George Allen all bought lobsters and shipped them out.

Lloyd built his own gear in the early 1960s, fishing out of Wallace and selling to Kennedy. There were almost forty boats fishing from Wallace at that time and a 400 to 500 pound daily catch was a normal. In the 1970s, you were lucky if you got a hundred pounds and soon the number of boats out of Wallace shrank to about twenty-five. Lloyd fished there until 1980 and quit, ironically, just as about the lobsters were coming back.

Lloyd loved fishing and has an abiding love of the water.

Merrill Jamieson

Merrill was born in 1937. He started fishing in 1953 at the age of sixteen as a helper for Earl Jamieson, who fished from Wallace for Ken Kennedy in a company gear. They fished the waters between Oak Island and the Stonehouse. Pugwash boats would fish as far as the Stonehouse, from the opposite direction.

Merrill fished only three or four seasons with Earl before he and his father took a boat from George Allen, who had a lobster stand on the Gulf Shore, where the Scottish Pines is today. In about 1957 Merrill and his father started to fish for Conlew, at "the Brook," near where the Felderhof's home now stands. The Conlew stand was the farther west of the two fishing camps along the half kilometre stretch of shore at Robinson's Brook. Winston Allen and Charles VanBuskirk owned the camp between Conlew's and the Brook itself.

In the early sixties, when fishing was done in an open boat, Earl Jamieson and Phillip Oxley were out in a very thick fog. It was early June and they were lost. They passed a salt boat leaving Pugwash and looming very large out of the fog. They yelled to attract attention, but went unheard and the boat sped on. After hours and hours of traveling and long after their lunch had been consumed, and their gas was running low, they saw a fellow lobster boat, and thinking that they had found their bearings at last they trailed it for a while before pulling alongside. Much to their surprise, they found that they were just off the eastern tip of Prince Edward Island! Their newfound acquaintance took them home, gave them a hot meal, and put them up for the night. Back in Wallace, their worried loved ones received a telephone call. The next day most of them gathered with most of the town to meet Earl and Phillip at the wharf — as much to praise their courage as to poke fun at their seamanship.

Merrill bought his gear from Conlew in 1960 and continued to fish out of Robinson's Brook for another three years and sell his catch to the company. One time as Merrill rowed his dory to his boat, a baby seal followed, resting its head against the dory. Thinking little of it Merrill, jumped in his boat and went out to tend his pots. When he returned he found the motherless seal with his head still up against the dory. This habit of following and clinging to the dory went on for several days, the seal even flapped ashore in attempts to follow Merrill. After several days, Merrill picked the foundling up in his arms and took it home to his wife, Sylvia. She called the people at the Shubenacadie animal farm and they advised her on the feeding habits of baby seals: the cubs don't suck the milk from their mother but rather push against her to force the milk to squirt out. Sylvia fed it some baby formula and eventually gave the seal to the Shubenacadie people, who took good care of it.

In 1963 Merrill began working year-around at the Pugwash salt mine, but he continued to fish during the spring season. He left the Gulf Shore and went back to fishing from Wallace, where he sold to George Allen and then to Phillip Elliott, who bought Allen out. Merrill dropped his gear size back to ninety traps and took out a class B license, reserved for fishermen who had year around employment and had been registered in the lobster fishery for an extended period of time.

One time Merrill received a call at the salt mine from a fellow fisherman concerned that strong spring winds were creating high seas and threatening to sweep his boat ashore. Together with his shift foreman, George Myers, he left work and drove down to Wallace to save his boat. They hopped in a dory and made for the boat. Just as they reached the larger boat, their dory swamped, and they were cast into the water. Luckily though, they already had a grip on the washboards of the lobster boat and managed to pull themselves aboard. Merrill and George prevented the boat from sweeping ashore, but other men saw their boats wrecked. The two miners had to remain on the *Cape Islander* for several hours before the winds died down, wet, shivering, but happy for the boat's rescue.

Fishing regulations are not created by law so much as by federal Department of Fisheries policy. Over the years these policies have dealt with such things as: designating lobster fishing as a limited entry industry, implementing a license buy-back program, restricting license transfers, establishing discrete seasons, limiting trap numbers and size, and defining legal carapace size and lobster escapement mechanisms. These policies are subject to change to keep pace with the often rapidly moving fishing environment and industry. When lobsters became very scarce in the 1970s, the department introduced three categories of lobster licenses to help reduce the number of outstanding licenses and eliminate a category of fishermen informally known as "moonlighters."

The three categories were outlined as stipulated in a letter from Romeo LeBlanc, Minister of Fisheries, in November 1976: 1) Category "A" — reserved for people who do not have year round employment outside the industry, nor any full-time seasonal job coinciding with the lobster season. These people may fish 100% of the authorized number of traps for their lobster district.

2) Category "B" — reserved for those who have year around employment or whose seasonal job coincides with the lobster season and who have fished consistently from 1967. These people may fish 30% of the maximum traps for their district as long as they remain in the fishery and may not transfer their license.

3) Category "C" — reserved for those who have year-round employment or whose seasonal job coincide with the lobster season but who acquired their license after 1968. These people

may fish thirty percent of the maximum number of traps for their district but may renew their license for only two more years until the end of the 1978 season and may not transfer their license.

In the mid-1980s, as the lobster catches improved, Merrill and six other fishermen, including Bill Smith, organized a petition signed by many people to have their licenses upgraded to Category A. Merrill and the other fishermen were turned down primarily because: only fifty-eight Category B licenses were outstanding in District 7B1 and District 8 out of a total of 1929. Such small numbers did not warrant a policy change given that the large majority of fishermen did not support them. As a direct consequence, when Merrill retired from the mine, he had to buy a Category A license in order to fish with a full gear. He bought his license from Paul Canfield in Wallace, who would only sell provided Merrill purchased his boat and traps. Merrill merged the two gears and sold the surplus boat.

In another example of the obstacles created by policy in those years, a fisherman who had sold his license as part of the buy-back program in 1979 bought a Category A license when the lobsters returned in the mid-1980s. The problem was that he purchased it in Big Island and transferred it to Pugwash. Many local fishermen objected because policy had traditionally prevented port-to-port transfer. Unknown to the local fishermen, however, this policy conflicted with the *Charter of Rights and Freedoms* and was struck down. The policies of the fishery sometimes required the probing mind of a Philadelphia lawyer!

Merrill loves the challenge, suspense and competition that accompany fishing for lobsters. Sylvia, his wife and lifetime companion, says that as soon as end of December arrives Merrill begins knitting new trap heads, eagerly anticipating the next season and the hunt for the elusive lobster within a sometimes cold and blustery sea. In 2003, Merrill celebrated fifty years of continuous fishing. All of his children, two daughters and a son, have fished with him and used their earnings to pay for their education.

Kennedy Family

Clarence Kennedy

The original Kennedy homestead was just beyond where the road to Port Howe turns towards Oxford but before the Rockley cemetery, where Murray Smith now lives. Clarence left the family farm in the early 1900s and started a lobstering business. At his peak, he had a factory on the Gulf Shore, a large one on Saddle Island and, it is believed, one in New Brunswick.

Clarence bought the Saddle Island operation from George Elliott. Then, when the Malagash salt mine closed and sold its buildings on the wharf, Clarence moved his operations off the island. All of Clarence's equipment, traps, and rope was stored conveniently in the old salt warehouse. One night in March the warehouse caught fire destroying all its contents. Clarence and his fishermen worked long hours to replace everything, even going so far as to buy old traps from around the country and patch them. By the start of the spring season on May 1, Clarence and his fishermen were back up and running.

In time, Clarence turned the business over to his son, Ken.

Ken Kennedy

Ken Kennedy was born in Pugwash on February 11, 1919. When Ken was 12 or 14, a fire in Pugwash burned him seriously leaving him unable to raise his arm fully and disqualifying him for military service in World War II.

Ken was hired as an engineer aboard the RCMP boat charged with tracking rum runners. Clarence owned and leased the boat to the Mounties. It had a powerful Fairmile motor. While Ken was away chasing rum runners off St. Pierre & Miquelon, his parents moved to Bayhead.

Ken's older brother ran a lobster factory in Aulds Cove down near the Canso Causeway.

Ken, befitting a good Scot, was said to be thrifty in his use of words and was not given to telling long stories or anecdotes, particularly about his long history in the industry. Ken passed his lobstering to his son Charles. He died in 2003.

Charles Kennedy

Charles was born in Wallace in 1957 and first began fishing after high school when he was seventeen years old. He fished from Wallace for two years, 1973 to 1975 with his own gear and sold his lobsters to his father, Ken Kennedy. Charles remembers that lobsters went "dry" for seventeen years, from the mid 1960s to the early 1980s. They returned, then peaked in the early 1990s, and have decreased since. He left the industry at the first downturn and he pursued his real love, flying.

Even though he became a pilot, Charles never lost his connections to the sea. He began helping his aging father with the lobster business in 1982 and gradually assumed a leadership role as the years advanced. By that time, the family had centralized their buying and holding operations in Wallace and Malagash. The Wallace plant is located on the original site of the very first lobster factory built in Wallace in the late 1800s. They purchased lobsters from Barrachois, Malagash, Wallace, Pugwash, Northport, and Murray Corner, New Brunswick. Charles' chief competition came from Chase's lobster pound in Port Howe and Elliott's in Wallace

Charles sold the Kennedy family firm in 2001 to Sky Fish Ltd., a Truro company controlled by a large Newfoundland company. Sky Fish failed to meet the conditions of sale, and Charles got the company back on April 22, 2003. At full capacity the processing plant employs about seventy-five people from Tatamagouche to Pugwash. The plant sells frozen and live lobster and ships to Boston via a trucking company.

The plant's lobsters come from an extensive area but primarily from among the thirty boats out of Wallace and the fifteen from Malagash.

Charles believes that the decline in the number of lobster factories was attributable to the drop off of lobsters and a tightening of health regulations. For example wooden floors and ladles were forbidden under tighter rules.

Lawson Smith

Lawson Smith was born in 1920 and comes from a family of fishery workers — he, his grandmother, and his father all worked in the lobster factories. He started at a factory when he was still too young to work in a boat, and didn't go out in a boat until he was eighteen years of age. Fred Brownell, whose father was called Honest Ira, threw three boys and Lawson overboard one day. It was sink or swim. Luckily, Lawson learned to swim.

A man had to be strong and tough to go. Lobstering was a way of life: when you graduated from school, you graduated to the lobster industry. Lawson went to grade eleven, choosing to work rather than travel to Oxford for grade twelve.

Lawson has a picture taken in 1915 of the workers at the Chambers and McKay lobster factory on Sand Point in Colchester County. His father, William (or W.C. as everyone called him), was the boss and wore a white shirt and tie and at times even had cufflinks. He is at the centre of the photograph, sporting a moustache. W. C. ran the Colchester factory for many years but it did not survive. Within fifteen years of the photograph being taken, many factories disappeared. The only thing left of the Horton's Point factory is a well pipe that is now below the high-tide mark: it gushed water for many years after the factory was gone.

Fishermen were paid by the pound, often this was four to six cents. Man Trenholm, who bought the Horton Point factory from Josh Allen and ran it until 1938, paid four cents in 1938 when he was undergoing financial challenges, but others paid more. Some fishermen, but very few, owned their own boats and fishing gear. More often, the boss fed the fishermen, put them up, and paid for their gear. Josh Allen, then Man Trenholm, owned all the gear that fished from Horton's Point.

Although as a rule Man Trenholm never told the truth, he was a good man who would give his shirt off his back to people. He was not educated so Lawson wrote his letters and read them to him. He was a great boat builder and taught the craft later in his life. He eventually learned to read and write and had a blacksmith shop.

Man would come to the factory on the 17th of March, and work alone for about two weeks, and then he would hire two workers to help him get all the gear ready. Each boat had 400 traps so, with eight boats, they had to take out 3200 traps and pile each gear separately. Each of the traps had to be inspected for fish and shells as rats would work their way through headings to get to these, thus destroying many traps. The paint on the boats would have worn off from rope hauling, so Man and his team had to scrape down and repaint the boats and let them seal before placing them in the water in early May. At the end of the season, boats were hauled up, motors taken out, and everything put in the boathouse with the traps. The boathouses were large but only eight feet high.

Man was full of hellery. For example, he would talk to the radio and demand that the man inside come out and fight him. Another time, at Elm Lodge owned by Man's wife Liz, the cook had a helper who was not washing. She complained to Man who shut off all the water. When the helper got up in the morning and claimed that she had washed he knew she was lying and fired her. Man Trenholm had a brother who was deaf and mute, but fished out of a lobster boat. While he could not hear, he could feel the rhythm of the motor and knew when it was working properly.

The fishermen left for the fishing grounds at four o'clock in the morning, a time when there is just enough light to see one's way out of the harbour. They would go out eight miles to the Trenholm Shoal, fish all day, and sometimes not return until ten o'clock at night. The fishermen who fished near the shore would get back sooner, as they didn't have to fight as much wind, but generally the men preferred working away from the shore. A boat could easily be pushed up onto the rocks in minutes if its motor broke down. No one went to bed until every boat was safe on shore.

The Trenholm Shoal was out of sight of land, so the boats traveled there by using a compass and a sounding lead to find the shelf. Only Man Trenholm's boat had a chart. Once the shoal was found in the spring, a marker was placed at its west end. In the early part of the season, the men who fished inshore caught more lobsters but the fishermen on the Trenholm Shoal would catch up by the end of the season. The shoal was about three miles long and twelve fathoms or seventy-two feet deep at low tide. Once you went off the rock it got much deeper.

At the end of the day the boats docked on a floating platform and the men transferred their lobsters to crates on the platform. The crates, when full, often weighed up to 150 pounds on the scales. A man from the factory weighed the lobsters at the platform. He had a small dory, called the factory boat, which he used to bring the lobsters to shore. Other factory men helped him bring the crates up to the factory dock, which was level with the factory floor.

There were about fifteen people working in the factory: a boss, someone to fire the boiler, someone to run the stationary engines, one for sealing the cans (called a sealer), and one that ran the well, two or three other men and about ten unmarried Acadian women. The factory had to be very clean, as it was regularly inspected. The box that held the lobster shells was disinfected daily with lime. Galvanized metal was used on the ceiling, walls and floor. Waste on the sloped floor was washed out to sea. Every day the walls were washed down by the men. The women would go to their quarters and change their clothes and come back and wash their working clothes in the steam boiler.

The men carried the lobster crates into the factory and dumped the lobsters into the boiler. The top of the boiler was lifted by block and tackle. The upright boiler was six foot by four foot by two foot, and the steam was carefully managed so as not to blow apart the lobsters. The lobsters were brought to a boil and carefully timed to cook for about twenty minutes.

A man scooped the cooked lobsters from the boiler with a net, and put them onto the cooling table. After the lobsters cooled, usually two fellows broke off the claws and tails and put them in separate containers. The claws would be sent up one side of the factory and tails the other. Another man would crack the claws but an Acadian factory woman would clean the meat out of them. Another man pulled the tails and a girl would clean the intestine and throw the meat into the sink. Another girl would wash the meat and then send it to the packing table. Lobster knuckles contained fine meat but they were very dangerous to clean and prepare. Girls' hands became very sore and inevitably, someone would get blood poisoning from the spines of the lobster. There was rarely time to extract meat from the lobster's body and legs; everyone was already working nights and days. Three weeks into a season, as the lobsters began to slack off, the workers would begin to pack tomalè. It was top quality and probably sold locally. Some factories, especially after the catches declined, placed the legs in an ordinary clothes wringer to extract the meat. This poor meat went with the fine meat.

Another man hauled the shells away for fertilizer. He came with a horse and wagon and a big square box and took the shells from the back end of the factory. When Lawson was young, he and his friends would climb up on the wagon as it passed their houses, toss out the bodies, and later extract the tomalè for their school lunches.

At the factory, the meat was placed in little wooden boxes and dumped into three piles on the packing table. There were two girls at the packing table who were very good at their job. One put tails in the bottom of the can and then spread fine meat on top; the other then put in the claws in an interlocking way so as to look very nice and to get exactly one-quarter of a pound. Then Lawson, who was usually let out of school for the first three weeks of the season, folded in a can liner and passed the can and cover to another man, the sealer, who used a mechanized sealing instrument.

The sealer then sent the cans to the steam retort room, which was partitioned from the rest of the factory. The vessel was about four feet in diameter. The cover was bolted down solidly and the steam turned on gradually. Up to 200 cans were placed on trays at once and stacked in the steamer. A block and tackle hung from the ceiling was used to remove the cans. The pressure was fifteen pounds and the heat was 238°F. The lobsters were boiled for one hour, and then the steam was shut down, the drain opened and the turnbuckles loosened. The operator had to be careful not to stand too close and he wore a pair of leather gloves and gauntlets to protect him from the steam. He lifted the lid and swung it sideways.

There were two alarm clocks on the shelf by the steam retort. One was a dummy and the other worked. The men set the time on the dummy at which the cans should be removed and when the two clocks read the same the batch came out. They never depended on the alarm clock itself. The cans were removed and dumped in the can room. All night long you could hear the swollen can covers retracting to their original shape as they cooled.

The girls worked very hard and every day confronted a big pile of lobsters. When they had a slack day or day in which the wind blew very hard, the girls had time to label the cans and put them in wooden cases holding ninety-six cans each. The fishermen never wanted to see a windy day but the girls always looked forward to it.

In 1931, in addition to their board, the Acadian girls filling the cans got $25 a month and those not filling cans received $20 a month. They came from Bouctouche and St. Edwards and could speak English to some extent. The two girls filling the cans, Mary Elaine and Pauline, who had been coming for many years, were very skilled and could speak English reasonably well. 1931 was the biggest year ever and from eight boats they packed 800 cases of lobsters — ninety-six cans at one quarter pounds apiece. All of the lobsters were shipped to England.

Mary Elaine was the boss of the women, hiring and looking after them. A priest would come down and inspect where they would be staying. The girls were bunked above the cook house, away from the fishermen in separate bunkhouses. The sign over the fishermen's door said "Trenholm Liars." The men working in the factory lived in another small building, the "crab shack."

On windy days, some of the guys would break out the liquor. If there were three windy days, sometimes the guys would come into the factory. However, the factory's name was a good one — the fishermen usually worked such long days that there was little time to get into trouble. Some other factories had a very poor name.

None of the Acadian girls at Horton's point married locally. The girls would come to Lawson's house on Sunday, when the factory closed, to sing and play the organ and help themselves to lilac flowers. They were dressed in bright colours and were a good group with no scandals. On some Sundays they walked five miles to and from to the Wallace Ridge church.

At that time Josh Allen was running a factory in Pugwash. Times were tough, but he would grubstake many people over the winter.

Most years were peaceful but in the later years fishermen ran lines across each other's lines forcing a man to lift twice the number of traps. There was a lobster war on Saddle Island in the mid-30s. The New Brunswick people came down and were going to drive the locals off.

Horton's Point operated until 1939, when war was declared. After the war other people came, including people from New Brunswick, and they fished 600 traps — 300 one day and 300 another, and cleaned out the lobsters.

Lawson married in 1948 and the same year he built a cold-pack factory for George Allen at Horton's Point. The lobsters were placed in big flat one-pound cans and the covers were pressed on but not sealed, and then shipped cold to Boston. There were about five boats at that factory. George didn't last too long as he was kind of the erratic. The factory became a fishing station and Lawson believes operations eventually moved to Shediac.

Man Trenholm lost his good name by "borrowing" gear, which gave him his first start. He had a very poor reputation. In about 1920, he built a new boat and left it out on the shoal with poor Clint Miller to fish from every day. He would pick him up in the evening time. A person from New Brunswick found the boat, claiming it was left adrift. Man started a lawsuit, which was not unusual for him, since he often had several of them on the go at any one time. He eventually got the boat back because he had built it and knew it and could describe it in detail.

Lawson never saw Man do an unlawful thing. He was all business in the boat; there was no fooling. Nevertheless, Man had had several close calls. He claimed to have swum ashore on three occasions but still never had a towline on him.

Pugwash

(Gilbert) Bruce Allan

Bruce was born on a farm on March 25th, 1911 in Bayside near Port Elgin, New Brunswick.

In the late 1920s Bruce's father, Frank, passed his lobstering operations to Bruce and his brother Max. Max limped from having had TB as a youngster. This left Bruce to do the heavy work and Max to do and supervise the factory work. They sold the factory in about 1937 and Bruce went into the military shortly thereafter for a period of five years.

The Allan brothers had about nine boats at the factory. They provisioned the boats and supplied the fishermen with all accoutrements, including room and board. Other, independent fishermen landed at the factory but their numbers differed from year-to-year depending upon the strength of competition from other factories and from the omnipresent smack boats.

The personal predilections of the fishermen, the size of the catch, the price of lobsters, the strength of competition and the shifting volume of demand were other factors affecting the number of fishermen selling to a particular factory. Sometimes the packer, in a big season, would not be able to process all the lobsters he received. Early storage methods, which consisted generally of piling the lobsters inside a building for several days, could not withstand oppressively hot spring and summer temperatures. In such cases, lobsters would spoil or die within hours. When the haul was unexpectedly high, everyone had to work around the clock to get the lobsters canned. They always tried to can the lobsters within twenty-four hours of their arrival on the shore. There were quiet days in the factory as well, as when there was a storm or the catch was light.

As with most packers, Bruce and Max hired their female factory workers from the Bouctouche area in New Brunswick. They tried to get the same girls every year as those experienced in placing lobster parts in cans were particularly valued by the factory. There was also a special talent involved in getting the meat out of the lobsters, particularly the soft-shelled ones.

After the cans were filled and sealed, they would be boiled for three hours, then drawn out, cooled marginally, and punched with a small a hole to allow the steam to escape in a rush. The hole was quickly resealed and the can made ready for shipment. They discontinued this method of sealing cans in the mid-1920s when the factory began to use a specialized steam and pressure controlled tank.

In the early days, the factory girls would come to Pugwash by train[1] and be brought to the factory by horse and cart. Later the girls were brought by truck. Over the years, a special bond grew between the Allan brothers and the seven or eight factory girls that worked for them. They would often visit between seasons and bring their immediate families. The girls bunked in a building near the factory. Naturally, the local boys would try to meet the girls, but Frank and Max made sure that they were not harassed. Nonetheless, the brothers felt free to visit girls in other factories and often tried. In one instance, they were driven back by rivals who pelted them with salt fish used for bait.

Some seasons were very good. Other years the fishermen might be lucky to come ashore with just two crates of lobsters. Bruce estimated that there were factories along the Gulf Shore about every two miles. Cape Cliff was at the end of the Gulf Shore stream of factories and it was a good place to tie up the area's small boats as it had several nice gullies. The rings that were driven into the rocky shore for this purpose can still be seen.

The competition among factories was very strong. The McInnis factory was about three-quarters of a mile from the Allan brothers' and each kept a sharp eye on the activity of their rivals' boats and when a competing factory "steamed up" to boil lobsters.

Bruce felt that the decline of small and local lobster factories began in the late 1940s as the original owners died or lost interest and had no heirs or no interested partners to take over. Many of these original owners, usually packers, were initially partially funded by wholesalers out of places such as Boston and Halifax.

Bruce didn't remember the well known packer from Antigonish, John Doyle, as having any lobster business in Pugwash. He remembered him instead as the owner of the waterfront store behind which his father, Frank, had built his smack and as the very jolly man who built the home in which his daughter still lives (beside Braden's store.)

Bruce Allen passed away in early 2004.

Note

1. At that time, Pugwash had two trains that arrived daily, one in the morning and one in the afternoon.

Clifford Allen

Clifford was born on November 7, 1933, and started fishing at thirteen years old as a helper for Ira Brownell in the fall and fished for his father in the spring. A year later, Josh Allen (no relation) rented him a small boat and about 250 traps, paid all of his expenses, and gave him eight to ten cents a pound for his catch. Clifford fished from Robinson's Brook, staying there six days a week during the season and with his father in town on Saturday night.

About 125 boats each with 400-600 traps fished between Cape Cliff and Pugwash in the forties. There were six boats out of Robinson's Brook for almost 4000 traps. George Allen had a stand just above Robinson's Brook (near where Dr. and Hazel Felderhof now live) and, above that, Gordon McInnis had his, a little beyond and towards Pugwash, and there was a third stand just below the Brook. Gordon Bollong was another fisherman who was down nearby, and kept his boat with the others in the boat harbour at Cape Cliff.

Fishermen assembled at the various stands around April 15 to get the gear ready and tended to bring their traps in around July 2, just when the lobster catch became poor in numbers and size. At the end of the season Josh gave his top fisherman $100 and a suit of clothes. Clifford received this award on several occasions.

When Clifford was just starting out, he had to use a grapple many times to feel out the ocean bottom and to learn where the rocks and the sand were located. One technique used in grappling was to put grease on the rope so as to have the sand stick to it as hard sand and rock will often feel the same to a beginner. Clifford remembers how the daily use of the grappling rope left a notch on the boat's side near the motor. Josh mistook the grappling notches for those left by the pulling of traps and, thinking Clifford confused in his fishing habits, asked him where he was pulling his traps from, the front or the rear.

The boat crew took the lobsters from the pulled traps and placed them in wooden crates that were dumped overboard when the boats came closer to shore. Olaf Olsen picked up the crates and transported them by horse and wagon to the shore where they would be moved to Josh's factory in Pugwash. If he couldn't wade out to the crates the fisherman would get in a dory and row them further ashore.

The men usually rose at 4:00 or 4:30 in the morning and within ten minutes or so after breakfast, were fishing. The food at the stand was the very best: bologna with eggs for breakfast, and boiled food for dinner. The men always had a big meal when they came in from fishing at 3pm. They would go to bed around 9pm, except for those nights they would be out drinking: bootleggers were very active in the forties and fifties.

The fishing groups came from Pugwash, Gulf Shore, and Wallace, and everyone got along well. Clifford can never remember an argument among these groups. The men got along better in those days and there was no animosity like there is today. No one ever mentioned the amount of lobsters they caught, as this was thought to be private information, and doing so saved a lot of conflict. At one time, some fishermen, including Clifford, fished two seasons by "loaning" their gear to another fisherman and, officially, at least, attending them as a helper.

Clifford remembers when two boats would bring in sixty to sixty-five crates, more than Olaf could take in his wagon in one haul. Out of, say, 3600 pounds, only about 300–400 pounds would be markets. Today, the reverse is true because human interference has caused the depletion of canner volume and numbers. In Clifford's opinion the lobsters will not return in volume until there is an upsurge in the number of canners.

Clifford bought his first boat in 1953. At that time a license was twenty-five cents. He left fishing in 1994, after forty-eight years of fishing! At that time he had a forty-four-foot boat equipped with all of the most modern of electronics — no need for grappling.

Clifford passed away in late 2003.

Joshua Reid Allen

Josh Allen was eight years old when his parents, Chesley and Frances, along with others in the extended family, came down to Northport from Anderson's Settlement, New Brunswick, in 1910. It was a lobstering family, as many local families still are.

Ches Allen was one of the first area lobster packers and an early owner of the Acadia Hotel. He fished and owned a small factory, where he packed lobsters by soldering the covers on cans. Ches died when he was just forty-nine years of age. His wife Frances married George Van Ember some years later. George later became Josh's boat builder. The hip roofed building he used, where Josh built many of his traps, is still there on Queen Street.

Josh met and married Ethel Chase from the Pugwash area on February 6, 1924. Ethel's father, Benjamin Chase, had a factory in West Pugwash and lived in the square house on the north side of the road near today's Camp Pagweak.

Josh began building and buying lobster factories in the late 1930s. His first factory was in Port Howe and his second was just up the shore from Crescent Beach near Frank Allan's lobster factory by then part of the Eaton Estate. Josh also had a fishing stand on the Upper Gulf Shore near where MacLean Drive is today. Ethel helped to cook and serve the meals for Josh's fishermen.

Josh's Port Howe factory was a good size for its time and had both a cook and bunkhouse. The operation was just by the bridge between the road and the water. The lobsters were offloaded at the wharf and transported by truck to the factory. They were then weighed and measured at the weighing station and sorted according to size, canners or markets. Josh's daughters, Reta and Ruby, and their school pals often helped line and label the lobster cans. The lobster cans were loaded onto a truck bound for Halifax and then shipped overseas through Josh's broker, a Mr. Sexton. Each can was stamped with the J. R. Allen logo, and one of the two brands, Maple Leaf or Evangeline. Reta designed the lobster on the can, and the image was eventually used by many other lobster entrepreneurs.

Josh bought the Acadia Hotel from his uncle in the 1930s and it became the epicenter of his operation for many years. The Hotel housed the offices from which his daughter Reta and Lila Demings oversaw the financial operations of Josh's ever-expanding entrepreneurial energies and was home to some of the fishermen, helpers, factory workers and cooks from Josh's nearby factories. Reta remembers hearing the workers, both men and women, playing the violin and tapping their feet from the other side of the hotel. At that time Federal legislation forbade a factory located in a spring season location from processing the fall catch and vice versa. Josh was successful in having this changed so that he was able to catch lobsters in both seasons but have just the one canning operation.

By the early 1940s Josh controlled a large but scattered operation consisting of a factory, several stands, numerous woodlots, a farm, a garage, a smelt operation, a hardware store and the hotel. He now had six boats at a stand on the lower Gulf Shore at Robinson's Brook near where the Felderhof house now stands, and counted Del Nicholson, Elvie Nicholson and Cyrus Allen among his fishermen there.

Josh leased an Irving Oil garage on the corner of Water & Queen streets where the parking lot for the Mundle funeral home is now. The garage was used primarily to overhaul and maintain his many boat engines and eventually those of his twenty vehicles as well as to fill them all with marked gas. He never sold his fuel. Fishermen's costs were posted by Reta and traced.

In the winter when the lobster season was closed, Josh packed smelts out of the old Salvation Army building. The fish were dumped on the freezer floor and just as they started to freeze they would be cleaned, glazed in water, put on trays and labeled No. 1, No. 2 or extras with the J. R. Allen label. Josh got into farming in order to feed all of his workers. His cattle provided the steaks and hamburger, and the leftovers from the many dinners he served to his pigs. Six separate cookhouses served his workers.

Josh and his son, Winston, stocked the farm with pure-bred shorthorns, with many traded and bought from Cyrus Eaton. His woodlots supplied his mill on Queen Street, which in turn supplied his factories and cookhouses. His hardware store near the Baptist Church was the main conduit for supplying his growing operations. The store was open to the public and occupied an old house that burned down in the 1960s or early 1970s.

Partly in response to the growth of his scattered operations, Josh built a much larger Pugwash factory in 1941. It had a green hip-roof and was located on the peninsula near the Palmerston graveyard as you head west out of the village.

About this time the industry switched from canning lobster meat to preparing cold packs and shipping live lobsters. Cold packs, both those with plug and window tops, were packed into freezer barrels (130 packs per barrel) and delivered in freezer trucks to Josh's brokers in Boston, Washington, New York, Philadelphia, Chicago, Toronto and elsewhere. It was said that Josh invented the cold pack. The can was lined with parchment, the tails were placed in the bottom, broken meat was put in next, then the claws were interlaced on top, the parchment was folded together and the can sealed with a plug top or transparent window top that allowed you to see the meat inside. Josh also invented little machines with brass wringers that the factory girls used to squeeze the meat out of the lobster legs. There might be up to eight girls at a time on these machines.

Live lobsters were sent to Boston in two separate trucks. There were no refrigerator trucks in the early days so the live lobsters were kept cold by putting ice on them. The ice came from Fraser MacDonald's pond, near where Fred Brownell lives today, and was kept in an icehouse by the main lobster factory. The people packing the lobsters had to be very careful that the fresh water from the ice didn't melt on the lobsters and kill them.

Lobster paste, or tomalè, was still canned the traditional way. Josh had a reputation for providing very good tomalè. He did two things differently from his competitors: he mixed his tomalè with a little lean meat and didn't douse his lobsters with cold water after he cooked them. This made his tomalè more solid than others and it was in great demand. Many schoolchildren had sandwiches day after day made with his tomalè.

Josh hauled water from his spring in Pugwash to his green factory, but he was always anxious to have a water source directly on the property. Every time he dug for a well, however, he came up with salty water. He gave up. But then one day in the 1950s he saw a sign from a Halifax company that all but promised him water if he hired them to drill. He did but, instead of water, they ran into a seam of salt. He promptly spread the news that he had found enough salt to last a hundred years, thus stimulating the subsequent exploratory work that led to the development of the Windsor salt mine in the village.

In 1954 Josh built a much larger factory, now part of the Seagull Pewter property at the marina. Both times Josh expanded his factory he reduced his costs below the competition, which further enabled him to expand his operations and become the top lobster packer on the north shore of Nova Scotia. By the 1950s, Josh was purchasing and bringing as much as a million pounds of lobster each year to Pugwash for processing and shipping.

Josh used mostly his own gear and fishermen but also bought from independent lobstermen: Fraser Pauley had a fishing stand in Shemogue and his wife, Lizzie, cooked their catch; Lester Turner, Wendell Baker's father-in-law; Charlie Elliott; and Murray Smith. Josh also bought lobsters from smaller packers such as Ken Kennedy in Malagash and Wallace, Gordon McInnis on the Gulf Shore, and everyone from Shemogue to Caribou to Pictou. And, of course, Josh pulled in lobsters in the spring from his factory in Pugwash, his stands on the Gulf Shore at Caribou and at Bayfield, and in the fall from his stands in Shemogue and Northport.

In order to keep such a large number of lobsters fresh, Josh maintained large twenty-foot by twenty-foot by twelve-foot floats just off the factory and one at the federal wharf. He also had a pound house by the factory. Sometimes, the lobsters were so numerous that Josh had to use cold storage plants in Moncton and Amherst in addition to all his storage facilities in Pugwash. The old Salvation Army hall on Queen Street in Pugwash also doubled as cold storage for Josh's lobsters.

The mainstay of the factory were the Acadian girls who came in from Richibucto, New Brunswick, and stayed for two months in the spring and fall. Every second weekend some might go home from Saturday at noon or from whenever the work was done until Sunday night. Those who stayed would have turkey dinner. The girls were fed well but were paid modestly from about $3.50 to $5.00 a day. There was no time clock. They worked until everything was caught up from the regular breaking up of the lobsters to sorting them into containers.

In the old factory, the girls would eat and sleep on the second floor. By the time of the last white plant there was as many as eighty Acadian women and ten men working in the factory. Such large numbers occasioned the building of new separate eating quarters. As well, Josh built a residence for the ten or so older women who were about sixty years of age and who did not enjoy the dancing and excitement of the younger women. In the evening time, the young ladies partook in dancing, fiddle playing and the amusements of the local gentlemen.

To break up the routine in the factories, many of the people played tricks. There was an Acadian man who was a very good artist and he was always drawing pictures, including many in which he captured some of Harvey Laird's unique supervisory behaviours. Harvey was, in essence, Josh's second-in-command and oversaw everything in the factory, including the retail counter. Gordon Benjamin was also an important figure at the factory as he looked after the vital and often dangerous job of keeping the boilers at the right temperature.

Josh was known and liked by Pugwash's most illustrious offspring, Cyrus Eaton. In August of 1955, Eaton was entertaining world famous intellectuals at his residence in Pugwash. *Newsweek* captured one of their encounters:

> **Scholar vs. Townsman:** In a gathering of men with strongly held opinions, the only conflict was between a scholar and the townsman. The townsman, Joshua R. Allen, a prosperous lobster man, arrived during cocktails bearing a box of freshly boiled lobsters that the intellectuals were to have for dinner with a 1952 Liebfraumilch. Eaton introduced Allen as "the lobster King of Pugwash."
>
> Allen told the party how to tell a male lobster from a female. Then he began explaining the mating and reproductive processes — in plain English. Julian Huxley kept interrupting, trying to explain things in scientific — and more polite — language. For a while, the two contested for the floor, but in the end the grandson of Thomas Huxley, the great naturalist, was routed by the Lobster King of Pugwash. Host Eaton restored decorum by recalling the dinner-table conversation about lobsters in "The Way of All Flesh."

Josh never fished. He packed for forty years, eighty seasons in total. Josh, like other packers, provided and looked after the people who work for him. Many fishermen stayed in his houses. He would give them advances against the lobsters and smelts that they would catch later on. He drove a hard bargain but could be very understanding as well. He never held a grudge. You could argue with him and call him names, and do good business with him the very day next.

One example of his approach was at a time when one of his friends had a brother who died in California. The man owed Josh $1000 and said he could not go because he didn't have any money. Josh canceled the loan and paid his fare to California.

But he was more than a business entrepreneur. His restless mind was always seeking new solutions, and he shared his inventions with all who wanted them. In addition to the clear covered cold pack can and the wringer for getting the meat from lobster legs, Josh invented the Allen bearing which was placed in a boat's motor to disengage it from the hauler when it wasn't in operation.

Josh never talked much about what he did at work, even with his own family or employees. Josh was a one-man band and kept all of the information he knew to himself. He had good men but insisted on overseeing everything. He was very hard on himself. His form of relaxation was to look over his cattle or take guests on boat rides, including Cyrus Eaton's Thinkers, but he never had a hobby.

Overburdened by all the challenges he faced, he had a nervous breakdown in 1958. His doctor gave him a choice "you could die a hero or live like a man." Given this option, Josh sold his business in 1958 to Conlew when he was fifty-six years of age. Conlew was an amalgamation of two other packers' operations and their names, A. J. Connolly and R. J. Lewis. Within a couple of years of selling his business, Josh passed away.

Reta recalls that for the most part Josh did well. She does not remember a year of loss, but then she never saw the final tally of revenues and expenses. Josh was probably the only person, other than the auditors, who had seen the income and balance sheets.

Ainsley Allen, Josh's eldest son and Reta's brother, played a large role in the operation of the factory. He started learning the business from the ground up in about 1950. He tended the retail counter, did much of the heavy lifting of the lobster crates, oversaw the cooking of the lobsters, was the time keeper, and reorganized and realigned the many buildings that had grown up around the factory. In 1958 when the business was sold, he became the foreman for the Conlew operation for the first year and then ceded the position to Leroy Leadbetter. Conlew eventually closed the Pugwash plant and moved its residual activities to Shediac. After a short time in Ontario, Ainsley returned to Nova Scotia in 1966 and became manager of the National Seafood Products operation in Pugwash. All the lobsters purchased in the area were shipped to the company plant in Cap Pele' New Brunswick. He worked with them until 1968, when he went to work for the federal government in Amherst and Halifax. Ainsley died in 1979.

Lobsters were landed at the dock, measured and weighed and then placed in either the floats to the left of the dock, as pictured above, or in another set of floats to the left of the Federal dock. In some instances, the lobsters were transported directly from the dock to the factory via the railroad trolley. Here they would be cooked and packed or sold at retail.

Reta Allen

Reta was born on October 30, 1926, in her maternal grandparent's house in West Pugwash. Her father, Josh Allen, was born in Anderson Settlement, New Brunswick, and lived in Port Elgin and Northport before coming to Pugwash. He was one of many Allens, including his father, who had moved east. Josh met his future wife, Ethel Reta Chase, in the Pugwash area and they lived temporarily with her parents early in their marriage. Ethel's father, Benjamin Chase, had a lobster factory near where Camp Pagweak is today.

After living in West Pugwash, Josh moved his family to the Acadia Hotel at the corner of Pugwash's Queen and Water Streets. The hotel was originally owned by Josh's father, Ches who left it to his son Burt, who in turn sold it to Josh. Reta's brother Ainsley and her sister Ruby were born in the hotel overlooking the harbour. McGregor Hunter built a house for Josh at 30 Queen Street and the family lived there for many years before eventually moving into 25 Queen Street.

Reta went to the Pugwash School. As children, Ruby and Reta worked around the factory, just as their younger brother Winston did in later years. Among other things, they placed liners in lobster and paste cans (before cold packs came into use).

Reta went to Oxford to finish her Grade eleven and then, at her father's insistence, she went to Mount Allison University to study music. She stayed there for only three weeks.

Reta returned to Pugwash in 1947 and went to work in her father's office in the Acadia Hotel. She and Lila Demings did the bookkeeping. Lila had worked there since 1939 and had some business training in bookkeeping and stenography from a school in North Attleboro, Massachusetts. Reta learned by working alongside Lila until she was able to do more and more.

When Reta first learned to drive, she and Ruby would go and pick up some of the Acadian staff from St. Edwards, Bouctouche, and Richibucto. Sometimes they needed to make several trips. Eventually the number of staff increased so much that Josh arranged to have them brought down by truck. Josh used local people, too, in his packing operations, but the universal finding was that the Acadians seemed to be better adjusted to work in the factories. A man in the Richibucto village, Adeodat Aresenault, nicknamed John by Josh, acted as middleman in obtaining Acadian staff.

Josh built a new office just up the street from the Acadia Hotel, now a private home next to where Orie (McCrae) Skidmore now lives. He also constituted his first company, J. R. Allen Ltd. Previously he had run his company under his own name and Reta remembers the huge numbers of signatures and authorities that were now required by the incorporation.

Reta and Lila ordered all the boat equipment as well as all the food for the cookhouses. Their purchases came from all over the Maritimes. Every year net heads and maple spears for the traps had to be purchased. Trap rings had to be ordered from Lou Tom, a supplier from the aboriginal community. Boats had to be repaired or built. The cook and bunkhouses had to be repaired and upgraded. Bait had to be ordered, and on location the icehouse had to be stocked, and on and on. There was a huge number of things to track, including a wide variety of fishing gear, purchases from fishermen, wages and unemployment insurance figures, transportation costs, cans and cases of lobster, and destinations where sales took place. The money streams for payments to both men and suppliers were in the millions

— a lot of money even today.

Reta looked after the payroll and her calculations included such things as tax deductions, unemployment insurance, and worker's compensation. Reta was often kept busy closing the books at the end of a season. Some fishermen would obtain loans from Josh in the fall, and Reta had to post all the specifics to the fishermen's accounts. There were charge books for gloves, boots, and oil clothing. At the end of the fishing season, each fisherman received a detailed account and a calendar from J. R. Allen Ltd. Reta was never able to complete a total payment for supplies or for the fishermen until the end of the season when all the numbers were in. She went to the Bank by car.

In the winter, Reta and Lila prepared United States export papers for the next season as this was a time consuming effort. They caught up on journal and ledger work and prepared for the onslaught of the spring fishing season.

Lila was a great and honest worker who worked six days a week during the fishing season. It was an eight to five job but as the business grew, she had to work more and more hours. Eventually, Josh had so much trust in her that he gave her power of attorney. While Josh was undoubtedly the main decision maker, when pressed, Lila and Reta would make decisions on their own.

The Acadia Hotel was sold to Conlew (Conley and Lewis) in 1958 and later to Donald Mundle who tore it down and built a garden. Lila left J.R. Allen Ltd. in 1959 and went to work in the office of the Cumberland Home until the 1970s.

The peak of the business was in the 40s. Most of the years were good: some ups and downs but nothing serious.

Winston Allen

Winston was born in 1941, the youngest of Josh Allen's children consisting of Reta, Ruby, Ainsley and Lola. He entered the lobster business in 1958 with his brother-in-law, Charles Van Buskirk.

Winston and Charles bought the old Alf Trenholm shop down on the lower Gulf Shore, just down the road from Josh's own shop and near where the Felderhof's home is today. The purchase consisted of land and a twenty-foot by forty-foot two-storey building. Downstairs there was a desk in the corner for Winston, a long dining table around which the men sat, a sink to wash up in, a storeroom, a stove and the cook's bedroom. Upstairs there were sleeping bunks for the men.

Winston and Charles ran six boats from their shop. One was purchased from Ainslie McLeod and the other five built by McPherson in Wallace. Each boat cost about $1400. Clifford Allen, Bill Allen, Herbert Ferdinand, Lester Myers, Tom Pye, and Jim Pye skippered the boats but Winston and Charles owned them and the gear. Winston and Charles incurred all expenses. Oftentimes they would extend credit to their fishermen to help them over the winter or through hard times. The men were paid about eight cents a pound for the lobsters they brought in, but the amount paid depended upon what other packers were paying their fishermen. Any difference between what was owed to the fishermen and what they in turn owed to Winston and Charles was adjusted at the end of the season.

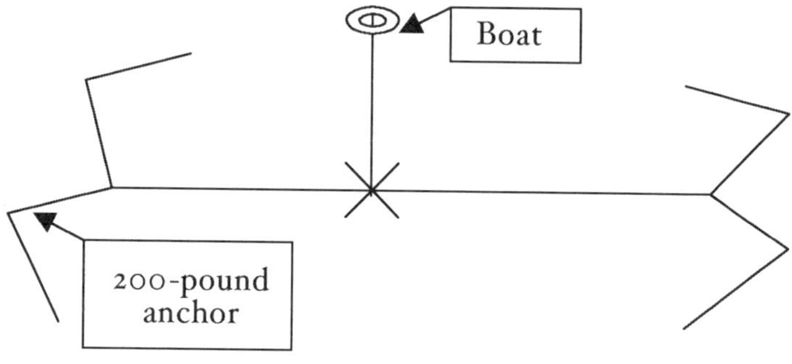

The boats were held fast to a sandbar in front of the shop by a unique combination of anchors. Each boat had a spring line attached to a standard 200-pound anchor allowing the boat to rise and fall with the tides and waves. Sometimes, to avoid relying on the tides to access the sandbar, the men would anchor the boats further out beyond the sandbar. The anchors were built by Royal Simpson in Oxford.

A large number of different people built the traps but, Hope Pye, Tom's wife, knit most of the heads.

The men tried to be out on their boats by 4 am if possible, then would come ashore in the mid afternoon, wash up, have dinner, and then take a snooze. They would wake later in the early evening, have a snack, and then go to bed so that they could get up again the next day at 3 am for breakfast. Mrs. Jenny Mahoney, the cook, would beat on the stovepipe to wake the men up. She had a bunk downstairs and at one time had two small children living with her. Mrs. Beatrice Stevens later replaced her. They were both excellent cooks. Altogether, with Charles, Winston, helpers, the cook, and the two children there were up to eighteen people eating four meals a day at the shop.

At the end of the day, the lobster would be brought ashore put in a truck, and sent to E. J. Bourque Limited in Cap Pele (Cape Ball). Typically, Winston and Charles would receive thirty-six cents for canners and thirty-eight cents for markets.

Charles, skipped his own boat out of Malagash and had two

helpers because his gear, at times comprising 800 or more traps, was so large. He used to sell to Clarence Kennedy but, after he and Winston bought the Trenholm shop, he brought all of his catch up to the Gulf Shore. He continued to fish in Malagash because he had always fished there and knew the grounds well.

Heavy ice in the Northumberland Strait delayed the start of the 1962 season until May 27. They had tried earlier in the month to tie ten knockers of traps (consisting of 10 traps per knocker) between two permanent ice poles. But on May 26 the wind shifted so they were forced to bring in all of the traps and pull their boats up on the shore. The ice was so thick that there was no open water to be seen. Despite the late start, the fishing was very good that season. The ice had kept the water cold and even in July the lobsters still had hard shells.

On the final day of that season, July 10, the seven boats, including Charles's, brought in 6000 pounds of lobster.

Another year the boats were caught in a huge storm. One of Herb Ferdinand's anchors broke loose and his forty-two-foot square stern boat broke free. The wind and waves drove the boat into Charlie Robinson's field and smashed it to smithereens. Another of the men, Lester M. Myers, said he knew the ground and water and was going to sail his boat to the Pugwash Harbour. Everyone thought he would never make it because of the dangerous conditions, and Winston even called his father to say he expected that Lester would never arrive. But, Lester had said, "See here, I am going to make it." Against all expectations he did!

Winston and Charles left the shop in 1966, just as the lobsters were failing. Charles later worked at the Windsor salt mine in Pugwash before taking early retirement due to illness and Winston went on to become a successful undertaker in Oxford.

Charles passed away in 2003.

Wendell Baker

Wendell was born in Marie Joseph in Guysborough County in 1933. He married Geraldine Turner and fished for her father in 1952 and again in 1957 and 1958 on the Gulf Shore. Geraldine's father fished for Man Trenholm and after that for McInnis. In 1959, the Bakers moved to Pugwash and shortly after Wendell had his own boat and gear.

As Wendell recalls, as one traveled from Pugwash along the Gulf Shore, Alf Trenholm had a lobster factory near where Harlow Hollis has his house today, opposite and just before the golf course. Then came the McInnis factory, near where the old church was located, and Stromberg had a lobster factory about half a kilometer past the Seventh-Day Adventist Camp and just before where the camp grounds are located today. James and John Ross had another lobster factory near the Stonehouse. The Robinson Brook factory was further down, on the Cape Cliff road and was the only one in existence by 1952. Much of the land previously occupied by the lobster factories is now under the waters of the Northumberland Strait.

Wendell remembers that lobster factories were more a phenomenon of the 1930s than the 1920s when money was difficult to procure. Many of the factories that arose in the 1930s were attributable to the slow, incremental accumulation of capital by entrepreneurial fishermen who passed their legacy to their children.

Other lobster packers had backing from companies like

Paturel who would extend some start-up capital. The processing part of the industry in Pugwash was phased out in the late 1940s with the exception of J. R. Allen's. Josh ran his operation until about 1960 before selling out to Paturel, later National Seafood in Shediac.

Josh Allen, the largest packer in the region for many years, had a frugal side. There was a factory inspector who came and looked things over to ensure that the factory complied with regulations. He had a habit of taking a bit of lobster as he went about his rounds and eating it. Josh heard about this and kept an eye on him and when he saw the inspector take a piece of lobster he went up to him and insisted that he a pay a dollar for the meat. This led to an intensive argument in which Josh said that if he did not get paid he would lock the inspector out. Eventually Josh got his money. George Allen, Josh's brother, was excitable. One time he tried to telephone the central office but was unsuccessful. After many tries he tore the phone off the wall, took it down to the central office, and while cranking it said: "I bet I can reach you now."

Factories really came into their own in the 1930s as technological advances made lobsters easier to can and ship live market-sized lobsters to large cities, particularly Boston. Wendell remembers when he was only seven years old, helping his grandfather separate markets from canners in Marie Joseph. In the 1930s, ninety percent of the catch was comprised of canners. With the start of the Second World War demand for lobsters increased. The fishermen, in response, took lobsters that were very small and even lobsters with eggs on them. The government stepped in just before the war ended and imposed size regulations and forbade the taking of berried lobsters. Without this act of conservation, the industry might have disappeared.

Girls either from the Eastern Shore or from New Brunswick did most of the factory work. The wages were ridiculous. They weren't paid until the end of the season and would have to buy small items "on tick." At the end of the season, a girl might go home with as little as five dollars. They worked a ten-hour day for fifty cents plus room and board. The conditions were not very sanitary: outdoor toilets, no hot and cold water for showers, etc. Yet the girls seemed to enjoy the adventure and the parties on some evenings, particularly Saturday nights. Some of the factory girls married local fishermen.

In general, the packers for whom the men fished were honest but if you contracted for five cents a pound you were bound to it, even if others received fifty-five or sixty cents. The packers were very insistent that a "deal was a deal." The top fishermen always got the highest price per pound.

In the 1930s and 1940s men and women came from places like Ecum Secum on the Eastern Shore to help with the lobster fishing on the Northumberland, where it was felt the pay was better. The men received a flat rate per pound and had only the expense of a helper or "his man." The helpers did most of the hard work, including hauling the traps hand over hand until the advent of the mechanical haulers in the mid-1940s.

The boats were kept small about twenty-six to twenty-eight foot long, to allow more stable hauling of the seventy-five to a hundred traps over the side of the boat. Fishermen also preferred pink stern boats because it made hauling long lines of traps easier. Wendell remembers that Art Daly owned the last pink stern in Pugwash in the early 1960s. Ira Brownell had previously owned it.

For many years, traps were connected and tied by manila rope. After a storm, the rope would sometimes become so chafed against the reefs that it would break and the traps would be lost. One reef was so hard on the rope that Wendell called it "heartbreak ridge" after its namesake in Korea. The manila rope was replaced by polypropylene in 1961. The new fibre was slow to catch on at first but, after the fishermen saw its advantages, they adopted it.

The good fishing grounds tended to be called after the people who first found them, for example: Chambers Grounds, Quereau Reef, Jackson Reef, and Hape Reef.

When Geraldine's father was fishing in the 1940s he received six cents a pound out of which he had to pay "his man." In 1952, the first year Wendell fished for his father-in-law, he was getting only thirteen cents a pound. The packing company paid room and board for the men and shouldered all of the expenses associated with the gear. The average catch was one to two hundred pounds a day and towards the end of the season the average catch dropped closer to forty or fifty pounds per day. It would not have been profitable for a single boat to fish and pack their catch but a packer with eight boats could make money. A can of lobsters at that time sold for ninety-nine cents.

When Wendell bought his own gear he sold his lobsters to Paturel, which supplied the bait. He fished from the Pugwash government wharf, but in the early 1960s, the local fishermen lobbied for a fishermen's wharf. The wharf was completed in 1963 or 1964. It was later enlarged in 1973 to accommodate fishing boats still tied at the main wharf and squeezing the ever larger salt boats.

All the really powerful storms "due northers," come from the north-northwest directly from Cape Tormentine. These are annual events, and most come at night when they are less strong and the lobster fishermen are at home. There are better marine forecasts today than in the decades gone by, and these storms are now easier to anticipate.

Wendell was out with the other fishermen one day and a wind came up from the northwest. They hurriedly made their way to their moorings but one of the boats missed and was caught on the sand. The boat's packer, McInnis, was deeply worried he'd lose the boat and with it a measure of his season's profit. Murray Smith ran for his tractor and with the help of 16 men, who held the boat upright, towed it to safety.

The worst storm Wendell can remember was in 1967 or 1968, when he was out in a new three-year-old boat. He had an old fellow in his sixties, Tom King, with him. Wendell was fishing and looked up and saw a black cloud in the northwest. He immediately turned his boat to shore. The wind came up suddenly and blew in squalls up to sixty and eighty miles per hour, taking the tops off the waves and blowing them horizontally. There were no cabins on the boats in those days and the single boat that had one had it blown off. Tom had to hold himself steady in the bow with a gaff. It took two hours to make a half-hour trip. One boat ended up on the Oak Island reef. Although there were many leaky boats, no one lost their life.

Everyone played tricks in those days and George Allen was often the butt of them — many played by his own men. But Man Trenholm was the biggest joker. Every year when a greenhorn arrived at the factory, Man would take him by dory to the boat on some pretext, say, to get some bait. On the return trip he would upset the dory and put both of them into the cold water. One time Thorpe Moody was working in the factory. Now Thorpe really loved his molasses and spread it thickly on his bread. Man put motor oil in with the molasses and placed the molasses strategically so that it was easy for Thorpe to spread it on his hot biscuits. A look of surprise came over his face upon the first bite, but being a proud man he persevered and finished the biscuit. That night he spent most of his time "in the can."

Jim Bateman

Jim was born on a farm on Scoudouc Road. After high school, he worked for four years in the local area with the Bank of Montreal. In 1955, he joined Paturel as its accountant.

Paturel was founded in 1936 by Emile Paturel from the St. Pierre & Miquelon. The story goes that since he didn't speak English, Emile Paturel decided to come to the Shediac area where he started a lobster plant at Pointe du Chene and another at Pointe Sapin. He operated these factories for a few years until he had a fire at Pointe du Chene and then moved to Cap Bimet, where the plant remains to this day. Emile Paturel sold his interest in the business in 1950 to managers R. J. Conley of St. Andrews and Allan Lewis from the Miramichi area.

In the mid 1950s, the company paid fishermen who owned their boats eighteen cents per pound for canners and twenty-five cents per pound for markets. Those who didn't own their own gear received twelve cents a pound plus their room and board as well as a flat fee for getting the gear ready. The cannery women were paid about fifty cents an hour plus room and board.

In 1957, Paturel (Conlew), purchased its first assets in Nova Scotia: the George Allen business at Horton's Point which consisted of a cannery, a buying station, a book house and bunkhouse and another buying station on the Gulf Shore. In those days, fishermen brought their catches ashore by dory on the Gulf Shore and the lobsters were taken to be canned at Horton's

Point. In 1957, the two formed an operating company named Conlew to run their activities in the Pugwash, Gulf Shore, Wallace, and Pictou areas.

In 1958, Paturel bought the J. R. Allen business from an ailing Josh Allen. Jim remembers Josh arriving at the plant on a Monday or Tuesday and saying, "I want to be out of this business by Friday." Paturel purchased the Pugwash-based assets including a cannery, cook and bunkhouses, the old Acadia Hotel (which was used as a bunkhouse), a filling station at the corner of Queen and Water Street, a boat building and trap shop also on Queen Street, an old Salvation Army church that Josh had converted to a freezer shop, a lobster storage location on King Street, an extensive property at the old brickyard, a buying station on the Gulf Shore at Robinson's Brook, a team of horses used on the Gulf Shore for unloading boats, some sixty boats with gear and a buying station in Northport.

Jim has two anecdotal memories of Josh that gives a colourful depiction of his mode of operation. At the time of the takeover, Jim was in the filling station on Queen and Water streets counting the inventory when Josh came in and asked: "What are you doing here young fellow?" Jim replied that he was doing an inventory count for the purpose of the sale. Josh said, "You're wasting your time. There is about $1,800 worth here." Jim countered, saying that it looked like about $1,200s worth. Josh replied; "Let's split the difference at $1,500 and you'll save a lot of time." It was a deal and they settled on that amount. Another time, after a few weeks of running Josh's old operation, Jim mentioned to Josh that the grocery bill for all of the cookhouses was very large. Josh remarked, "Look here young fellow, the cooks all make great chocolate cakes. Ask them to make them early and place them on the table before the meals. The men will come in, see them, and eat a big piece. You will find that your grocery bill will go down noticeably." And so it did!

Jim remembers other employees at Paturel. The foreman

of the cannery, Harvey Laird, had a great sense of humour. One time when Jim was visiting the plant with another foreman, there was a considerable amount of bubbling activity caused by the presence of a large number of sixteen-year-old girls. Harvey greeted his visitors by saying, "Welcome to the romper room." Margaret Jamieson was the managing accountant for a number of years and a very loyal and devoted employee who took a sincere interest in the company. Ainsley Allen, Josh's son, managed the plant for a year or so. When he left, Ainsley decided to get back into the lobster business and bought a property on the Gulf Shore. After a few years though, he sold his assets to Paturel — the second time Paturel bought out the Allens. Ainsley worked with the unemployment insurance office in Amherst and Halifax, and died young, just forty-eight years old. Other stalwarts were Les Allen and his wife Lola. He was the lobster buyer and a troubleshooter while she was the cook for many years.

In the spring of 1958, just after the purchase of the J. R. Allen operation, Paturel bought 465,000 pounds of lobsters. In a spring in the mid-1970s it bought just 27,000 pounds. The dramatic decline in the supply of lobsters caused Paturel to close both the J. R. Allen plant and the one at Horton's Point and consolidate their operations in Shediac. Jim was the general manager of the J. R. Allen cannery when it closed.

In about 1965, Paturel purchased the Gordon McInnis property on the Gulf Shore, which Mac Pittman now owns. There were about ten fishermen and a bunk and cook houses associated with the property. As well, the purchase included a horse and a high-wheel wagon for loading and unloading the lobster boats. McInnis had his office in the back room of the cookhouse and a fine home in Wallace on the hill opposite the fire hall. Paturel operated the property for ten to twelve years.

Most of the help at the plants, cooks, fishermen, and cannery helpers, came from outside Pugwash because the village's seasonal workforce was small. Many of the fishermen came from the Cape Tormentine, New Brunswick, or the south shore of Nova Scotia. The cannery people came predominantly from the Richibuctou village in New Brunswick. Jim believes that there were two primary reasons why the canneries were staffed by Acadian women: there was a large number of seasonal unemployed people in the fishing villages of southern New Brunswick, and it was very difficult to get staff from farming towns where people had no experience or inclination to work with sea products.

Paturel was sold to National Seafoods in 1965 and Conlew folded up at that time. National subsequently sold Paturel to Jim and John Bragg of Oxford in 1988. Ten years later they sold their interest to American Holdco from Boston. Jim is now vice-president of procurement.

When Jim first went into the business, the market for live lobsters was ninety percent to the United States. Today the United States is still very important but a lot goes by air to the United Kingdom, France, Germany, Belgium, Italy, and Spain as well as to Asia and Japan. Frozen lobster products have also found their way into many of these countries.

The Paturel business has grown steadily since its inception, from less than $1 million in sales in the mid 1950s to more than $100 million today. They continue to operate two lobster plants, one in Shediac and the other in Morell, PEI.

Amanda (Desroches) Benjamin

Amanda came from Bouctouche at the age of sixteen to work in J. R. Allen's cannery in 1937. Thirty other girls from Acadian villages like Richabucto and St. Edwards came with her on the back of a truck. She was encouraged to come by some friends who had worked at the cannery before and was accompanied by five of her sisters.

The cannery had two floors: the factory was on the bottom and the women's eating and sleeping quarters were upstairs where the girls had their own bunks. Their cook came from Richibucto. The men who worked at the cannery slept in a separate grouping of cabins. Amanda worked extracting meat from lobster knuckles, a dangerous job due to the "thorns" on them. One time she contracted blood poisoning after being pricked. Josh was the only boss Amanda could remember at the cannery; he controlled everything. He also brought lobsters from Cape Ball to his cannery for processing.

The cannery girls received lots of attention from the local boys. Amanda's sister was in love with a local man, a Mr. Daley, but their mother wouldn't let her marry him because he wouldn't become Catholic. On most Sundays the girls would go as a group to the local Catholic church.

Unbeknownst to Josh the girls hosted parties on the second floor on many Saturday nights, and there was usually a fiddler, an accordion player and someone on the guitar.

Amanda was the only cannery woman (other than Olive

(Bernard) Smith who left the factory to work as a maid) who married a local man. She met her future husband — Gordon Benjamin — early in her time at the factory. He worked sealing cans and looked after the lobster boilers. Amanda had always had her eye on him, even though he had another girlfriend and all the girls liked him. He would often take water from the boilers to the girls so they could wash their clothes. Amanda stayed at the factory for a month after most of the other girls had left for the winter and got to know Gordon better. Before she returned home she asked Gordon who he would choose to marry and he said, "You, of course." They were married in 1943. He was forty years old and she just twenty-two.

After they wed, Gordon and Amanda moved to a house close to the factory and they lived there for several years. But by the late 1940s they had three young children and needed a bigger space. Gordon and occasionally Amanda, too, still worked at the cannery and so Josh let them move into a cabin near the factory. Josh thought so well of Amanda that he wanted her firstborn to be named after him. Amanda couldn't do this, but did give her firstborn the middle name of Josh.

Amanda remembers that at the factory, as in any other workplace, there was a fair degree of poor behaviour. One of the Acadian fishermen would get up early in the morning and go out to fish. He always brought back lots of lobsters and Josh thought the fisherman was the very best he had ever seen. It was only many years later that it came out that the man had been stealing lobsters from the floats and passing them off as fresh catch.

Lloyd Brownell

Lloyd was born in Northport in 1920. Although Lloyd had polio when he was young and had an incapacitated leg as a result, it was never a great hindrance. He fished first with Ivan Polley in 1939 as a helper and was paid a dollar per day plus board. He was a helper to Hudson Trenholm the next season. He started work at Canada Car in Amherst in 1940 and stayed two or three years before returning to fishing.

In 1949, he bought his first boat for about $350 from Josh Allen. It was thirty-eight by twelve feet, built by Downey, and had a small gas engine for hauling attached to a "nigger's head." In about 1954, he put a shaft to the main motor with a power takeoff to haul traps. The boat lasted four years but was old when he got it. He bought a new one for $100 and it lasted eight years. The final two boats lasted about ten years and each cost about $25,000 without motors. In addition to the Downey boat, Garnet McPherson built two of Lloyd's boats, two came from New Brunswick.

Lloyd went to Lunenburg one spring to buy two motors at $5,000 apiece, one for him and one for his brother Fred. His last boat had a nice cabin and all the latest technology. Old boats without cabins were terrible. In one of his earlier boats, Lloyd built a small shelter/cabin from which he steered the boat.

Lloyd spent the winters building his gear. He fished the spring the first year out of Pugwash and then switched to the fall season for the rest of his fishing life. The fall season began in

August when the weather was good and lasted until the end of September when the weather began to deteriorate. You were permitted 300 traps in the spring and 250 in the fall.

In addition to Josh Allen's factory, Gordon McInnis had a factory in the late thirties down the Gulf Shore, just below where the golf course is now. Alf Trenholm had a fishing outfit where Harlow Hollis lives today. Charles Van Buskirk and Winston Allen had a place on the Gulf near Robinson's Brook. And Clarence Kennedy had a spring factory at Cape Cliff.

Lloyd had his own boat and gear but sold his catch predominantly to Josh Allen, later to National Seafoods and finally to Emerson and then to Earl Chase. At one time George Allen (Josh's brother) and Gordon Bollong had a smack boat and would come alongside the fishing boats to purchase a crate or two of lobsters. Most fishermen would sell small amounts to them to keep up an element of competition and to reap the benefits of a few extra cents per pound. In addition, "Kenny" Kennedy bought lobsters directly at the wharf.

In 1965 the lobsters left the Strait and didn't come back until 1983. Lloyd was one of the few that kept fishing during this time, even though he might only catch thirty to forty pounds every second day and the unit price was low. Warren Allen and Bob Allen were the only other fall fishermen to keep at it. Others kept their licenses but didn't resume fishing until the lobsters came back, which some how seemed unfair.

During these depressed times, the lobsters were scarcer in the fall than in the spring so more of the spring fishermen stuck it out. The spring grounds were larger and extended up to ten miles down the coast. To make ends meet during these years, Lloyd bartended at the Legion in the evening and drove his truck. In the spring, he would weigh lobsters brought in by Jim Ferdinand's fishermen for National Seafood and take them to Shediac for processing. National had a large floating web of crates at the north end of the main Pugwash wharf from which they operated. Ironically, the largest lobster he ever caught was in

1972 during the period of the scarcity. It weighted 13.5 lbs and had a nice hard shell and firm meat. He caught it in one of his two large traps.

Lloyd cannot remember any strong animosities or trouble-making among Pugwash fishermen, unlike in other areas. He stopped fishing in 1990.

Howard Ferdinand

Howard was born on March 27, 1941, in Pugwash. His brother, Herb, who is five years older, started the fishing tradition in his family.

Howard left school in the spring of Grade Eight and at age fourteen, and started fishing for wages as Hamp Van Ember's helper. He fished three seasons out of Pugwash as a helper; the first year with Van Ember, the next year with Lloyd Allen (who used Josh Allen's equipment), and the subsequent year with John Chase.

When Howard worked for Lloyd in 1956, he stayed in the Acadia Hotel. Everyone who fished for Josh out of Pugwash stayed there. Three men shared the room that Howard lived in, and there were about fifteen men in total in the entire hotel. It was a lot of fun. They had a good cook and everyone carried on. There were movies in town, the Tide's Inn was open and Ed MacKenzie and Minnie Lang's was a well-known hangout. In addition, some of the older guys would go over to the factory to meet the girls. Howard has vivid memories of some of the men and Acadian girls playing guitars and singing on the bank at the cemetery. Some of them were pretty good.

The year Howard worked for John Chase, the Federal wharf was overloaded with gear that belonged mostly to Josh Allen. The fishermen's wharf had yet to be built. Therefore, the crew ran a bunch of poles out in the water just to the north of where the fishermen's wharf is now and built their own temporary wharf for their traps and boat.

Howard remembers helping Jiggs Allen build a boat in the old brickyard building. Including the motor, Jiggs estimated his cost would be $600. Jiggs also built boats in Gordon Bollong's shed on Victoria Street. George Van Ember built boats in a shed on Queen Street for Josh Allen. Like the boats built by Jiggs, George's were towed by tractor with about ten men aside holding the boat upright through the main streets to the harbour. The last boats built in Pugwash were in the late fifties, although Garnet and Bob MacPherson in Wallace and the Kennedys in Malagash continued the craft for a few more years.

At the age of seventeen, Howard joined the second battalion of the Black Watch in Sussex, New Brunswick, and spent a three a year stint at Camp Gagetown. He returned to Pugwash in 1960, bought his own gear, and started fishing for himself in 1961. He purchased his equipment from Paturel, the company that had bought out Josh Allen and, as a result, owned most of the gear in the harbour.

In the 1960s a typical fisherman would have 600–700 traps, although some like Art Pipes in Northport had 1000. It would take a couple of days to get the gear out to the fishing grounds. To-day, with larger boats and only 300 traps each, the men can place the traps in two or three trips. In the early years, Howard sold his catch to Paturel and then to the Kennedys and Emerson Chase. Paturel later sold their boats and gear in Pugwash, and retreated to New Brunswick. National Seafoods then bought them out.

When Howard started fishing, the factories on the Gulf were no more; only Josh Allen's factory was left in Pugwash. Howard believes that the collapse of the small factories and of Josh's factory soon after it was sold to Paturel was a result of a better ability to ship lobsters live. When the factories were first closed, all of the area's lobsters were trucked to Shediac, but eventually that factory closed too. Now, of course, it is possible to have live lobsters in France within 14 hours of being landed on the dock in Pugwash.

Howard left the lobster business in about 1980 when the lobsters went dry. He sold his gear locally and his boat to someone in P.E.I. The federal government bought his license for $1500 as part of a program to reduce the number of outstanding licenses at that time. Attached to the sale, however, was a proviso that Howard could buy back the license within five years, which he did for $6500 in 1986, just as the lobsters were coming back. Howard fished every spring season until 2000. His routine was to fish and then go on construction. Howard's wife, Ella, started fishing with him in 1986 and stayed with him to 2000, a total of 14 years. She only missed six days in total over that time. Howard's sons, Stephen and Scott, got their start as helpers with their father and then bought their own gear. Both fish today.

Victor Matheson

Victor was born on the Gulf Shore in 1943 and went to the local school for nine years before going to the Pugwash District High School.

Vic's first year fishing was in 1960 with George Irving. There were eight boats fishing from McInnis's stakehouse on the Gulf Shore, near where the old Baptist Church had been. The boats were captained by Harold Jamieson, Rolley Refuse, Murray Smith, Charlie Elliott from Wallace, Wilford Turner and a Mr. Pye both from Ecum Secum and Thorpe Moody from Wallace. The stake house was located near a stream and close to where the traps were set. McInnis owned all of the boats and the gear and provided room and board for the men. Corn beef and hash was a popular meal. Breakfasts were hardy and lunches were prepared in a bag or in a personal lunch box. Prunes were served often for dessert.

Vic fished that first spring season and began preparing the boat and gear on April 15. When he went down to fish the first year there was six feet of snow on the side roads. The roads were like swamps. The fog was so thick some days the men had trouble seeing the landmarks by which they triangulated traps and had to use a compass.

He remembers his first time on the boat: the ice came in after the boats were out on the water but before most of the gear had been placed. The men were going to pull the boats into the shore but the ice had packed the boats in so instead they decided, at about four o'clock in the morning, to take them into Pugwash. They jumped into a slender Shelburne dory, pointed at both ends and about fifteen feet long, and rowed for forty-five minutes pushing aside ice cakes as they made their way to the boats. As the fishermen motored into Pugwash in single file, the seas were running high and at times Vic and George could not see the boats behind them. The waves pushed, heaved, and shook the boats. Vic became seasick. His captain, showing no sympathy, kept telling him to take some of the rotten bait to quiet his stomach. When they got to Pugwash, the harbour was blocked with ice and they were forced to turn around and go to Wallace. Twelve hours after they reached their boats, the men finally arrived in Wallace. Vic was a bit unsettled by the adventure as he could not swim a stroke. The other men took the dangerous trip in stride.

Vic remembers catching about 600 pounds of lobsters a day that year. The traps were smaller than they are today and there were about 10 traps a line. The men used a sounding device, a piece of lead on a rope, to find the reefs.

In the deep harbours the fishermen could go and come anytime they wanted, but when fishing from the McInnis stakehouse the men generally "ran the tide" leaving and returning to the stakehouse depending on the tide. When the tide was in, McInnis and two men would meet the boats and unload, weigh and put the lobsters into a truck to be sold elsewhere. When the tide was out, they took a horse and cart out to the boats. Eventually fishing from the shore, with its attendant challenges, disappeared as public infrastructure — roads, wharfs — improved.

George had a "hit and miss" or "make or break" Wisconsin five horsepower motor connected to the hauler. Eventually this was replaced with a hauler connected directly to the engine.

One of Vic's jobs when he arrived on shore in the evening was to use the high pump to fill the three five-gallon gas cans and take them, the bait and new crates by dory to the boat. On weekends, Vic sometimes took his buddies out for a few beers and a boat ride. Like all young helpers, Vic was adventurous and stayed out late at parties, slept little and was often very tired when the day's work was done. Many mornings he took a Gravol to help improve his disposition.

As Vic remembers, by 1960 there was only McInnis, who lived in Wallace, and the Robinson Brook stakehouses in operation. McInnis sold to Peturel in about 1963.

For Vic lobster fishing was a good learning experience. You had to work hard and had to be there on time regardless of whether you went to a party the night before or not. Many fishermen accepted that their young helpers would often burn the candle at both ends and were understanding, provided it had no impact on the work. Often, on warmer days, you would see the helpers stretched out sleeping on the bow of the boats as they made their way from the lobster grounds to the harbour.

Vic was paid $300 for his two and a half months of work, and worked for George for three seasons. Abandoning the life of a "homard man," Vic took his lobster earnings, bought a used car and left for Toronto.

Vic stayed in Toronto for a brief period and returned to study as a car mechanic and taught the trade for many years at the School for the Deaf in Amherst. He currently resides with his wife Vivienne on the family property near where he used to fish.

Hudson (Huddy) Trenholm

One day, the same year he left school, Huddy went down to his father's boat, which was kept on Alf Trenholm's property behind where the Legion now stands, and set out for a trip up the shore with his brother Sandy and a group of friends. They motored under the train bridge and saw something in the water — a body floating face down! Not knowing who it was, they decided to go to the captain of the Department of Fisheries Patrol boat to help them pull the body out of the water. Tom Allen was fishing nearby and said he hadn't seen any sign of life from the cutter captain. So Stuart Mallard ran to the Mountie while Huddy and his friends went back to the body. When they turned the body over, it was the captain! He had had a heart attack, slipped, fell overboard, and drowned.

Huddy left school on his sixteenth birthday, the same day he got a disappointing report card. He fished with his father in the falls of 1956 and 1957 and with Fred Brownell in the spring of 1956. His father, Hudson Sr., fished mostly what was called drop traps — five in a place — whereas Fred Brownell fished eight or more with a buoy at both ends.

Huddy fished 1958–1962 with his own gear. His boat was a Downey, built in New Brunswick. It had been his father's boat but when his father passed away, the boat was sold to Conlew. Huddy eventually bought it back for $2200. Huddy got his bait from Josh Allen who kept it in a shed in barrels. He also swept for gaspereau near the Red Banks in the Wallace River as well as for herring in Port Elgin.

Josh Allen was such an important figure in Pugwash that if there was no parking spot available at the Co-op he would park in the middle of the street and just go into the store. Josh was always good to Huddy and his father. One Christmas, Hudson Sr. was a bit scarce of money, but wanted to make it a good holiday for his large family. He saw Josh coming down the street and stopped him to ask for a loan of $400. Josh responded by saying that, given Hudson's large family, he should instead ask Lila Demmings, Josh's bookkeeper, for a cheque for $600. Hudson Sr. and Gordon Bollong were among the top fishermen in the area. They had spent their entire lives doing it and were very professional.

Most of Josh's lobsters were canned and sold to the States. He moved his lobsters from the marina over a rail track that crossed the highway and went right into the two-storey factory. The factory girls slept on the top floor. One night in the mid-fifties the boiler, which was housed in a separate building adjacent to the main factory, blew and blasted right through the roof like a rocket. No one was hurt, as the man tending it had just stepped out.

In the 1950s and 1960s, a full time Department of Fisheries cutter marked and monitored the fall/spring line from the Federal wharf. The cutter would travel along the line and wouldn't bother buoys that went fifty feet or so out of bounds. However, in about 1960, the fishermen began to take advantage of this and went further out. They were warned to pull them back and, while some did, others left their traps in place. The cutter pulled these up and smashed them leading to an altercation with the fishermen. A bunch of the men got "horned up" and threw gas on the cutter. But an eddy had opened up the space between the boat and the wharf and the fuel never hit the boat; instead it floated on the water. The fire traveled harmlessly downstream. Someone called the Mounties and names were taken, but everyone refused to testify and the people responsible got off.

After the 1962 season, and catches dropped, Huddy left the fishery. He sold his boat for $400 and gave his gear away but returned to the lobster industry in 1982 and fished with Gary Bollong until 1985. He bought his own gear in 1986 and fished with his helper Robbie McLaren from Pugwash Junction for fifteen more years.

Many changes had come to the industry in the twenty years he'd been away and since. His father fished with a compass and a watch to time the distance between traps and triangulated using landmarks on the shore. His father was so good at it that he could even work in the fog. Now, with better gear and electronics all you need to do is look at a map on the plotter. There are also fewer fishermen now than there used to be. At one time, there would have been forty-five to fifty boats fishing the two seasons out of Pugwash. Now there are less than thirty, partly because when the lobsters were bad, the fishery office said if you weren't using your gear you either had to sell your license back for $2000 or it would be cancelled. Art Daly and Sunshine Pollard were two fishermen who experienced this. Another change is that the full time cutter is gone. A small patrol boat checks the fall/spring line only two or three times a year; each fisherman has to remember where the line is. In addition, a fisherman must have a license to get trap tags.

Remembering Legendary Fishermen

Alf Trenholm

Alf, Hudson Senior's father, fished and packed out of Northport and was a good fisherman. He was from a well-known lobster fishing family in New Brunswick, but settled in Northport when he married Hilda Brownell, Fred Brownell's aunt. Hilda was the backbone of Alf's operation and took care of the books. Alf was very plainspoken in a disagreement. He and Hilda would get into some stiff arguments over business.

In the late 1930s, Alf Trenholm had a lobster factory in Northport and one on the Gulf Shore, and probably one in East Wallace. Hudson was the only son until seventeen years later when Shep Trenholm was born. He had two daughters both of whom married and moved away.

Mary Ann, Hudson's wife, does not believe that Alf ever fished. Mary visited some of the factories and believes that the canned lobsters were shipped overseas. Alf would send a truck to Richibucto for the factory girls and men who stayed in the factory. He ran a mixed arrangement owning some of the equipment while some of the fishermen owned their own. His cooks were primarily local. Alf was a good businessman who made lots of money. Mary remembers "a crafty old bird who was very plain spoken." Alf owned an old factory at Robinson's Brook, later selling it to Charles Van Buskirk and Winston Allen. Hudson and Mary would live there with their children in the summer and had many fond memories of the place.

Although Alf could not read or write (he could only draw his name), he ran a good lobster business for a long time. He was in the lumbering business as well, and could walk through a lot and nine times out of ten, provide an excellent estimate of its timber.

In later years, Alf loved his sailboat — possibly a hangover from his early fishing days. In the late 1950s he was sailing from

Newfoundland and went missing in a large storm causing a widely read search by both air and sea. Eventually Alf turned up — somewhat the worst from wear, but proud that his seamanship had trumped under circumstances that would have spelt the ending of a less competent sailor.

He loved to play cards; he was especially good at 45s. Alf also bought furs throughout the winter. One night a buyer came in and offered Alf $14,000, but The Alf said, "not enough," and took the furs away.

Alf was also a great hunter and would get two deer for every one that other people shot.

Both his son, Hudson, and his grandson, "Huddy," became respected fishermen in their own rights, perpetuating a fishing family that goes back more than 100 years.

Ebenezer (Eben) Cameron

Eben's grandfather emigrated from Scotland in 1800, settling first in Prince Edward Island and then in Port Howe. He built a small cabin there and underneath it dug a storage place for vegetables. Times were difficult and some neighbours were actually starving; one year people had to boat to River John to get potatoes to survive.

Eben's father was born in Port Howe, but after Eben was born in 1853, the family left for the United States. For many years, Eben's father drove an oxen-pulled iron wagon for a Boston bank.

Eben and his father returned to Nova Scotia in the 1870s and Eben started fishing lobsters in a small sailboat not much bigger than a dory. Then, in about 1880, he built a shed and partnered with a blacksmith from near Oxford. Building on what Eben had learned in Boston about how fishermen in British Columbia canned and soldered, the two men invented a unique lobster canning system. Soon Eben and his father were operating one of the first lobster factories in Cumberland County. When fishermen landed with their catches, factory workers met them with a horse (or oxen) and a two-wheeled dump cart. The work crew put the lobsters into jute bags, loaded the cart, and drove back to the factory.

There the lobsters were boiled in large galvanized tanks. Each had a canvas sealer between the rim and its cover. The factory women removed the meat from the lobsters and the shells were dumped in a box out back of the factory. From there the dump cart would take the shells and bodies to fertilize Eben's garden and feed his pigs. Meanwhile, the meat was canned. The central system consisted of two stove covers placed six inches off the floor and connected to another stove cover placed about three feet above them by an iron rod. This third stove cover had a lid around it into which you placed the can and using the soldering iron you would snap the lid on with one flip. One crew member, Johnny Moore, was famous for being able to process ten cans per minute!

The fifteen-member factory fished, cooked and packed the lobsters and were responsible for cutting the wood for the factory boilers. They used the multipurpose dump car to transport the wood from the forests, as well as to cart seaweed up from the shore to fertilize the Cameron land. The crew slept upstairs and ate downstairs.

After the turn of the century, in about 1910, Eben acquired the area's first single-cylinder motor boat which was very efficient and would, on calm days, pull the factory's fishing sailboats into the shore.

Many lobster factories grew up along the shore at this time. Eben and his father sold out to other interests and the factory was closed down. After several years, in about 1920, Silver Allan built a new factory and replaced Eben's soldering iron mechanism with a new sealing machine. Fred Dakin took over the factory after that. Clarence Kennedy had a factory on Donald Cameron's Beach, Josh Allen had another close to the Port Philip bridge, and there was another factory near Tony Bay. Silver Allen had another brother Gilbert, who also owned a factory.

Sometimes people would go back into the woods and can lobster for their own use over the winter months. The local sheriff, Eben's relative, who h ad heard about it, would stay away unless specifically directed to do something about it. He knew people needed to have cans of lobster for the winter. He intervened only once when he heard that some of the cans were being sold.

Smoking herring was a big business at that time. There were a number of smoke houses in Northport. The fish were placed on rods over the fires that were kept burning for as long as two and a half months. Then the fish were taken to the Pugwash wharf on scows pulled by sail or rowboats, and from there shipped to the West Indies. Roads were very poor at that time and the main transportation route was by water. Silver Allen had a big general store in Pugwash, near where the Cyrus Eaton Park is today, and Eben would go to the store by sailboat and tie up at the store's wharf.

Frank Allan

Frank Allan had a large farm in Bayside, New Brunswick, and a lumbering business in nearby Port Elgin. He also owned a sizable smoking business and would ship the herring out by the box load from Hardy's Station.

In 1922 Frank bought his brother's Pugwash lobstering business and built a lobster factory that still stands today on water Street and is commonly known as the Eaton Tearoom. Frank had had to tear down parts of a large existing building on the lot, and those parts he kept (four good rooms), he used to house his female factory workers for several years. During construction of this, his first permanent factory, Frank erected a makeshift building on lower Crescent Beach (where the swamp is today), in which to process and pack lobsters. He tore it down after that first season was over. Frank had about four boats which he provisioned. Several independent boat owners also sold their catches to him.

Ches Allen owned a small factory up the Crescent Beach a bit (on a lane that comes down from where Jack Allan lives today) accessible by a road that once ran along the beach. Ches passed the factory to his son, Josh. A great spring in the field behind the factory fed operations. Josh would later spread lobster shells as fertilizer on this same field.

Every lobster factory needed a steady and dependable flow of water and the "tea room" got its from a nearby overflow, near the high-water mark. There used to be many of these around, including one in the home originally built and occupied by Frank's older brother Silver that currently houses the Mundle Funeral Home.

Frank lived across the street from his factory, not far from where Silver had built, in a construction that consisted of three buildings joined together and siding King Street. The first two buildings were used as a place to house fishermen and a cookhouse. Frank and his family lived in the third building. Since this was before electricity came to Pugwash the house was lit by a combination of water, gas-air, and carbide, which fueled a triad of lamps through small tubes. The lamps were lit by standing on a stool or ladder.

In 1924, Frank sold his factory to Cyrus Eaton, who made the building into a tearoom. Bruce Allan, Frank's son, remembers that before Eaton bought and renovated Eaton Lodge, it was an old wreck inhabited by two families. A crew came in from New Glasgow to plaster and fix it up.

Frank maintained a stand in River John and he would truck lobsters up to his other factory on the Gulf Shore. In the spring, he could hardly get up the Gulf Shore Road as it had not been graveled and was covered with deep ruts. Even when it began to dry at the end of May, a good rain would turn it into a slippery mess.

Frank had a large smack boat built with a Studebaker engine powerful enough to allow the boat to ply its trade from River John to its home port of Pugwash. One time on its journey it ran out of fuel about where the Irving Cove is on the Gulf Shore Road. The boat proved an unprofitable venture as it consumed a great deal of gas and was sold in the mid-1930s.

Towards the end of the 1920s Frank was aging and turned most of the operation over to Bruce and his brother Max.

Gilbert and Silver Allan

Brothers Silver and Gilbert[1] Allan were active in the lumbering and lobster business in Pugwash.

Silver, in addition to his other lines of work, had a merchandise store on the Pugwash waterfront, which he sold in the early twenties to Ren Brownell from Northport. There was no electrical wiring in those days. When Brownell went down to the basement with an oil lamp to put the cat out, he slipped and fell, burning the entire store down.

Silver was very large and a gentleman at all times, except when faced with aggressors. He told his nephew Bruce that part of his business success lay in that he feared no man. As an example, Silver related that when he was transporting his lumber or mobile steam mill from place to place and anyone shouted at him or made to block his passage, he would step down from the wagon and fight, and always, it seemed, beat the other man.

Gilbert bought Balculm's[2] lobster factory, which was just to the East of where the current Seventh Day Adventist Camp is on the Gulf Shore Road. Gordon McInnis had the factory behind where the old United Church used to stand on the Gulf Shore Road and John Black had the next factory down towards Pugwash, near Matheson's Cove. Silver eventually sold his lobstering business to his much younger brother Frank.

Bruce remembered a gigantic storm in about December 1931 that washed a large boat up near the tearoom and flooded the road in front of Silver's home. By that time, however, a Dan Macarthur, the local stationmaster, owned the home. It now houses the Mundle Funeral Home.

Gilbert built and owned a large home near the corner of

Water and Black Street. It was destroyed by fire, dealing Gilbert a large personal blow.[3] He followed up with a brick house (still standing today on Black Street), thinking that it was less likely to catch fire.

Ira Brownell

Ira was born in 1890 and died in 1966. He fished all his life and was considered a very good fisherman. He started fishing from a 25-foot sailboat, and at that time, you needed to row a lot. When fishermen adopted the "one lunger," a single or double cylinder marine motor, he, like many others used it on his sailboat.

In Ira's day everyone sold their catches to the packers, many of whom were also active in the lumbering business. If you asked to borrow $100, they would suggest you take $300 in order to get you in to debt to them. Ira fished out of Northport in the fall and out of a Gulf Shore location in the spring. He worked for his brother-in-law, the Northport packer Alf Trenholm, until they got "horned up" one night and into a strong discussion.

Soon after, Ira moved to Saddle Island, about half a mile from Malagash Point, and fished for Tuttle King, who eventually sold out to Clarence Kennedy, grandfather to Charles. Those who fished from Saddle Island would get there by crossing with a team of horses at low tide. Ira later moved to Pugwash and fished for Josh Allen.

Ira's sons, Fred and Lloyd, fished with him in their youth. When Fred started fishing with Ira for two seasons out of Northport, Ira's eyes were bad and Fred had to spot the buoys. One time in 1935, Ira went from Northport to St. Pierre & Miquelon to get a load of rum and sugar for a rumrunner out of Port Elgin. Ira's lobster boat was caught in a storm and the sugar was soaked. He managed eventually to run his boat onto the Amherst Shore and walked back to Northport.

Fred also fished with Ira at Saddle Island. Lloyd began to fish with Ira at Saddle Island about 1937 when he was seventeen. Fred and Lloyd built Ira's gear. Lloyd remembers knitting heads for the traps when he was twelve years of age, shortly after his mother died and while he was still in school.

Notes

1. "In 1907, the Conservatives again fielded Smith and Webb as their candidates. The Liberals nominated Gilbert N. Allen and Alexander Hollis. Allen (1865-1948) was a New Brunswick native who had been in business at Pugwash for some years as a packer and exporter of canned and live lobster." (See History of Pugwash, page 262.)

2. Balculm came from a well known family from the Eastern Shore out of Halifax. They were known for being active in many ventures, including politics.

3. Around one o'clock on May 31, 1927, Mrs. Gilbert N. Allen discovered the second floor of her home in flames. Her husband, an ex-M.P.P., for Cumberland was working at one of his lobster factories on the Gulf Shore. The furniture on the first floor was saved but the second floor was lost. It had been originally built by H. G. Pineo, a former M.P.P. and had twenty-one rooms. One section of it was more than 100 years old. It was the third or fourth house in the Province of Nova Scotia to have hot water heating. Other reports said the house had more than thirty rooms, including a secret room. There was also a wine cellar. It was a land mark, not only in the Pugwash district but in the County of Cumberland. (See *History of Pugwash*, page 324).

Port Howe

Ross Cameron

Ross Cameron was born in Port Howe on Cameron Beach Road, on May 4, 1917. He went to Grade Four in school. He was one of five children, and times were difficult, especially after his father died. His mother remarried six years after his father's death and had six children with her new husband. There were a total of eleven children in the household.

At exam time in November of grade four Ross looked at his paper, tore it up, and left school forever. He went to work on farms and did chores and, in the spring of each year, there was always lots of work for him at the herring smoke houses. When he was fourteen, he went to the Annapolis Valley to pick apples and returned again the next fall. He was paid eight dollars a month plus board: a good salary in those days.

That second year on his way back home he got only as far as Parrsboro where a little freighter in port needed help and Ross was hired aboard for the same pay as he received for picking apples. Ross's first sea trip was to Boston. The captain was very good to Ross and took him into the city, which was quite an experience for a country lad. During that trip, he also cut his eyeteeth on the ship's rum, which came in five and ten-gallon kegs. The sailors were all sitting around drinking from tin cups, and Ross sat down to join them. He drank two cups of rum and was sick for two days. Soon after, he quit his job on the freighter and went back to working on farms.

In 1932, when Ross was fifteen, he went to fish lobsters.

Young people were expected to do a man's job in those days. Eben Cameron was the first man he fished for. Fishermen had small boats in those days; they were only about thirty foot long with single cylinder motors and didn't go any farther than two to three miles from land. The lobsters were plentiful but the price for them was low. Clarence Kennedy owned a lobster factory near Cameron's farm. Ross fished for Eben for a couple of seasons and then spent the rest of the year working on Eben's farms. He then farmed for Fred Irving down on the Gulf Shore at the same farm Fred's son Keith now owns.

Through family connections to the Allans and to the other Camerons, Ross fished until 1939 in the waters between Cape Tormentine and Saddle Island. He also fished for Hazen Trenholm for three years out of Cape Tormentine.

When war broke out, Ross went to Sussex, New Brunswick to join the army but did not pass because of a heart murmur. He left for Truro where he drove a bus and a taxi and also met and married his first wife in 1940. They had two children. After Ross married, he worked on the railroad until 1948, but managed to hold on to his pension service until 1954 by giving the break man one day's pay per month.

In 1948, Ross ran a public bus between Malagash and Amherst, and three years later started driving schoolchildren into Tatamagouche. When the regional school was built in Oxford, he won a contract to run all its school buses. In 1959, he sold his buses and went to California for a time with his wife and son. There he bought a piece of land for $40,000 and later sold it for $2,500,000. He left the United States soon after and came home.

He still had his house in Oxford and now built a cottage at Port Howe Beach. He remained restless so he bought the house next door and converted it to the Sandpiper restaurant. It was a good business, especially after Irving Oil set up a garage beside the restaurant and leased it to Ross. However, when his wife became ill, Ross realized he could not run the business alone. He sold the restaurant the same year.

Ross returned to fishing in 1961. He owned two boats, fished from Port Philip and sold his catch to Ken Kennedy, a packer who had gear out of Malagash, Pugwash and Port Philip.

In 1964, Ross sold his boats and went to school in Pictou, where he got his master's license. He took a job as a fishing officer on the *Maya*, a cutter based out of Pugwash on which his brother, Allison Cameron was the main skipper. There were three to five people working on the boat depending upon the nature of the work at hand.

As a fishing officer, he encountered all sorts of trouble. Locally, Ross and his crew would often have to chase boats with undersized lobsters for several miles. They would blast their horn and the chase was on! When they caught a boat though, there was seldom violence in the fracas. Ross and his men would take pictures of the offenders, their boat and their lobsters and would seize and count the catch. Other times, Ross would find many traps placed below the line that marked the fall and spring fishing territories. When the cutter crew caught these culprits, they would destroy the fishermen's expensive gear and fine them a modest one to two hundred dollars.

In the early 1960s, the federal government replaced the *Maya* with a new cutter named *5831*. The boat had been transported overland by freight from British Columbia to Pictou. It was powered by two 440 Chrysler Marine engines and registered at 39.9 knots. The skipper kept a 38 pistol on the boat and slept with it under his pillow, guarding against fishermen who would throw anything, even torches, on to the boat from the dock at night.

The cutter also saw a lot of trouble from the Indian reservations along the New Brunswick coast of the Gaspe Peninsula. One night in Richibucto near Big River Reserve, Ross and his men went up river to stop some illegal fishing. Things got very rough and Ross wrote many charges. Back at the dock at around six in the morning Ross heard an old car pull up. He looked out the porthole and saw a group of Indians. He told his guys to get ready for action and then went up to the car, pulled open the door, and asked the group what they wanted. The startled Indians said they had come to apologize for the night before. An astounding end to a very difficult night.

One of the more colourful packers that Ross heard stories about when he was a young man was John Doyle. John would do anything for fun, especially at his wood camp. One fellow placed his boots at the end of his bunk every night and at the crack of dawn, would jump into them and dance up and down yelling and singing to wake everyone up. The other men found this very annoying, so one night John nailed the boots to the floor and when the man jumped into them he fell to the floor amidst the laughter of all. On another occasion John went to the barn and got a scotch thistle and hid it under Rowley Elliott's blanket. Another story has Stan Smith from Pugwash send John a collect telegram in winter at his woodlot in Antigonish saying that they had lots of white stuff in Pugwash but they were not going to deliver it. John would have to come and get it himself.

Weldon Cameron

Weldon was born on February 21, 1912, in Port Howe at Cameron Beach in the same two story house where one of his sisters was also born. His father was Eben Cameron.

While at school, Weldon fished lobsters with his father, Eben. He had a room upstairs and often at night could hear and see rum running boats coming and going in the moonlight. One time when he was about fifteen years old, he saw a boat coming in during the night and heard a truck motor. He slipped on his clothes, hid in the moonlit bushes and saw men take kegs from the lobster boat and put them into the truck. There were some Allens around Cape Tormentine who made lots of money from running rum.

When he was eighteen years old, his sister married Charlie Elliott from Port Howe and Weldon went to fish with him in Pomquet, near Antigonish. Weldon was a strong and husky lad, and pulled about 350 traps every day hand over hand. Charlie fished and baited the traps. The first year Weldon worked for Charlie he was given wages. The second year he was given one-seventh of the catch and made $342 for two and a half months, more than most men made in a year. Weldon bought a car with some of the money he made. There were lots of lobsters there near Antigonish and Charlie made enough money fishing for himself to buy a home and to get himself started. Charlie moved to Wallace and fished twenty-seven years for a man in the village who lived near the big white church. Charlie always received a big checque at the end of the season because he was a good fisherman, didn't spend gas much and didn't lose any of his traps. Charlie and Weldon's sister had thirteen children, all of whom are still living. In fact, one of Charlie's sons, Philip, buys, boils, and sells lobsters from the Wallace wharf today.

Man Trenholm fished and had a factory in Lazy Bay, near Wallace. People tell a lot of stories about him. For example once there were two cows at a nearby farm and no one could figure out why one of them didn't give much milk. So the farmer and his dog stayed up late to watch the cow. And there came Man very early in the morning to take a bucket of milk for himself. The farmer set the dog on Man while he was milking, upsetting the bucket and giving Man a temporary bum foot.

After fishing with Charlie for two years, Weldon traded his old car for a tractor and harrow, and returned home to help with the family farm. The farm equipment cost $750 but now it would cost at least $28,000. Weldon worked in the woods for two winters for John Doyle, a man reputed to have caught more lobsters than any man who ever fished. He was often fishing after dark and many suspected that he fished below the line. John Doyle fished for Fred Dakin, but in the winter he and another man ran a wood camp of about twenty-five workers. John had a wicked sense of humor, always smiling and playing tricks. Once an old fellow was giving his horse a drink at the river between the sawmill and the bunkhouse. John had another fellow tap the horse yanking the poor man into the water. John had a great laugh and gave the soaked driver a new suit of clothes. John later moved to Marigonish, near Antigonish, and started a very successful factory in the sheltered harbour. He made lots of money and his original factory is still a big business today.

Weldon never went back fishing. He stayed at the farm for a year and a half, then moved to Truro where he worked for a man who put in sprinkler systems for six years. Later he went into the plumbing business for himself for the next fifty years.

Earl Chase

Earl was born in 1951 in Port Howe and went to the local school. At the age of five he went to the elementary school in Pugwash and graduated from Pugwash District High School in Grade Twelve.

Earl's life was tied closely with the fishing seasons. Every March during the school year, he would help his father paint buoys and then he'd fish his father's boat at the opening of the spring and fall lobster seasons. His father, Emerson, always fished his own gear. In the spring they would land off the old wharf in Port Howe but fish from Pugwash. In the fall the Chases always fished out of Port Howe. The law at that time stipulated that the same boat couldn't be used for both the spring and fall season so Emerson had two boats. This boat rule was changed in the late sixties when DFO introduced classed licenses.

Earl fished his own gear in 1964 and 1965 using a twelve-foot boat with a five horsepower outboard motor. He fished about thirty-five traps. The license for fishing was twenty-five cents at that time and he could fish all he wanted.

Emerson started as a smelt fisherman and got into the lobster business in 1953. His ambition from his first days in the industry was to sell as many lobsters as he could directly trap without selling to the factories. He fished from Port Howe and sold his catch on Saturdays and Sundays from a small woodshed on the wharf. In about 1954, Emerson's second year of fishing, he moved his retail operations to the present location and eventually was selling up to 20,000 pounds of lobster a year.

As Emerson's business grew he added a second boat to his assets, purchased a smack boat, and bought green lobsters from other fishermen at the wharf. Emerson and another retailer, Gordon Bollong, worked closely during these years doing deals together. Earl remembers being just tall enough, at the age of eight to read the retail scales used to weigh the lobsters. At that time, cooked lobsters sold for thirty-seven cents a pound; in March 2003, live lobsters sold for $13.75 a pound.

On Tuesdays and Thursdays Emerson traveled to Amherst and Moncton to sell lobsters directly to such places as the Lorna Doone Restaurant, the Fort Cumberland Hotel, the Colonial Inn and Goodwin's supermarket. His regular customers used to call him "Red Chase with the red lobsters."

Emerson's galvanized tin cooking tanks were made locally by Art Demings and at best lasted a season. During the winter, Emerson used the season's tanks as steam pots for preparing the bows employed in the making of lobster traps. The process of cooking lobsters in these tanks was simple: you boiled the water, added some salt, took the lobsters out and let them cool, sometimes overnight on a shelf. After that, you simply rinsed out the tub and you could begin again. To cook his lobsters today, Earl has to have stainless steel tubs, rinse them out very carefully, sanitize them, and then cook the lobsters at exactly 80°C and cool them to 4°C.

Even though Earl always worked during the summers and weekends selling lobsters, Emerson didn't let Earl cook lobsters until 1973, the year he passed away. Emerson took special pride in his careful cooking and in the quality of his cooked meat. Emerson had an old axe handle to stir the lobsters as they cooked: symbolically, it was his staff of authority. As Emerson left for his final trip to the hospital, he gave Earl a new axe handle, demonstrating in a way more powerful than words that Earl was now in charge of the business.

Earl took the business over in the fall of 1973. While Jean, his mother, knew the business thoroughly she never interfered, allowing Earl to lead despite the fact he was a rambunctious youth in his early twenties. Earl was a bit cavalier as he went overnight from being an employee to being the owner. He had recently married and said he would try the business for one year to see if he liked it. He made changes immediately to the operation because, being Earl, he had to do it his way and in the first three months he lost $7,000. Earl eliminated two of Emerson's lines of business: a fifty-acre crop of potatoes and buckwheat, and a gravel truck which, as a Liberal, he'd likely not get to use now that the Conservatives were in power. In addition, he expanded on the things he liked to do: lending money to fishermen to buy boats, gear, and motors and working cooperatively with the fishermen to obtain their right of first refusal on their lobsters. Typically, a fisherman would ask Earl to buy a truck, a lawnmower, or a sofa for his wife as part of a deal in which money and favours would be traded for lobsters. Earl was buying all of the lobsters in Pugwash, as well as Kennedy's lobsters out of Wallace and Malagash and from many other places along the shore including Lismore and Glace Bay.

Earl's early years in the business coincided with the collapse of lobster catches in the 1970s, and so he diversified into buying rabbits and selling squid, and expanding his father's smelt business. The rabbit business did very well bringing in 1200–2000 rabbits a month as far away as Dorchester and Port Elgin in New Brunswick and as far east as Malagash. He dressed the rabbits on Mondays and sold them to grocery stores such as IGA and Sobeys on Tuesdays. He bought them for $0.75–$1.25 a pair and sold them for $1.65–$2.15. The rabbit trade declined eventually for two reasons: as the prices for fur increased, trappers used rabbit meat for bait; and the Department of Health brought in new rules stipulating that rabbits had to be especially slaughtered, prepared, and inspected before they could be sold in grocery stores.

Between 1979 and 1981, a tremendous amount of squid

washed up on the local beaches. Earl obtained 15,000–25,000 pounds a day and trucked them to Shediac, where he sold them to the National Seafood Company for processing and use as bait in their business. At that time, there were still smelts in the local rivers but they were small in size and in quantity. Mysteriously, after the squids left, the abundance of smelts, which had been the source of much employment in the winter months for so many years, declined drastically.

The biggest years for smelts in recent memory were 1967 and 1970. In one net, on just one tide, Earl caught 14,000 pounds of fish, and in some years Earl and his father would get a half million pounds of smelt. In 1969, there were twenty-six nets on River Philip. Earl had his four nets, two abreast up and down river and monitored them by snowmobile, sometimes making as many as ten trips a night. At one time Earl employed three or four women to clean smelts. Smelts remained fairly strong in the local rivers and provided much employment in the winter months until the early 1980s, when their decline coincided mysteriously with the disappearance of the squid.

In 1975 a strange thing occurred. Earl had Bob and Warren Allen working for him and they "netted" 1200 pounds of Spanish onions and 800 pounds of carrots. Earl went to the Ministry of the Environment who told them that there were 250 people working upstream in Oxford and only twelve fishermen on the river, so he better go home and just forget it.

One November day in 1981, Earl was taking squid to National Seafood. One of the managers recognized him as someone who hauled lobsters back from Shediac to his retail operation in Pugwash. The manager predicted that lobsters would follow the squid and that it would not be long before there was such an abundance of lobsters in Pugwash and vicinity that Earl would have to haul lobsters back to Shediac because there would be too many for his retail operation to accommodate. Sure enough, in 1982 Lloyd Brownell and Bob Allen, among other local fishermen, were doing well and taking their catches to

Earl. Earl bought lobsters from five fishermen out of Little Shemogue, and in 1982 they had their best year yet. Earl remembers taking four crates of lobsters from one boat. The Little Shemogue harbour was afloat with lobster crates! Earl bought them all and was back for more the next morning. As an independent broker, Earl had licenses in both Nova Scotia and New Brunswick and sold for the best price.

In 1985, Earl converted his father's salt-bait shed at the Port Howe wharf to a cooking and tank house. The tank held one hundred crates of lobsters. That same year he bought land at Pugwash harbour and later, in 1992, he moved his tank house there. About that time, Earl knew that in New York and New Jersey fishermen kept their lobsters cool and in tanks fed by saltwater wells. He also knew that salt was seeping into the wells of both Fred and Lloyd Brownell, who lived near his property on the harbour. This gave Earl the idea that there might be a salt seam under his land and he began to dig for a salt well. At 220 feet he obtained ten percent salt in the water, the same as for ocean water and the temperature stayed cool at about 5°C both summer and winter. He connected the well outflow to his tanks and found that his lobsters kept fresh and lively for long periods of time. More recently Earl added a freezer to the well outflow lowering the water temperature even further. Earl saw two major advantages to having the lobsters in a cold water tank rather than on floats in the harbour: it protects the lobsters from flash freshwater flooding from the Pugwash River and it stores the lobsters colder, longer, keeping their shells harder for a longer period of time and making them more sellable.

Earl has transport companies do the hauling among the fishermen, his tank operation, and his distributors. In 1993, he started air-freighting lobsters to Europe. Today about half of his sales goes to Brussels and the other half to a broker in New Hampshire for distribution in the United States. He has about twelve people working for him eight months a year and just four people from January to April.

Jean Chase

Jean Benjamin was born in South Pugwash and went to the local school until Grade Eight. After that, she went to the Pugwash School and traveled by shank's mare. She was still too young when she graduated to enter college, so she taught school on the Gulf Shore, at Maccan, and in Amherst. In time, she became the principal at the new South Pugwash School (1959–1980).

When she married Emerson Chase in 1950 and moved to Port Howe, she became involved in the lobster business. Emerson's grandfather, Benjamin Chase, was a packer on the North Shore. A family picture shows five sailboats at a wharf, all of them owned by five brothers who fished for their father. Emerson fished lobsters for a few years and always fished smelts in the winter, but he was primarily a fish dealer. Emerson bought most of his lobsters from the five or six boats that fished from the wharf at the end of the old bridge in Port Howe. He also got fish from Pugwash, from his own packer boat, and from his smack boat in Port Howe. He called fishermen on a C.B. radio to arrange price and quantity.

Jean did everything there was to be done: she packed lobsters and smelts, worked at the retail in the summer and kept all of the books. Emerson opened the retail store in 1950 when the market lobsters sold for fifty-five cents per pound. Gordon Bollong also owned a retail outlet and did his own fishing, unlike Emerson, who eventually concentrated on being an intermediary in the trade. They were good friends and not very competitive with each other. Jean really disliked keeping the shop open until nine or ten o'clock in the evening, especially on Sunday. Customers got used to the late hours and would come very late: she remembers someone coming at two o'clock in the morning and buying just two small lobsters.

Most of Emerson's lobsters were sold locally or to supermarkets like Sobey's. He even sold to Ontario on many occasions, especially when people had special events and banquets. Still, the market for lobsters was not as strong as it is today.

Jean remembers one very difficult season when Vaughan Pauley brought his first day's catch home in his lunch can. On another occasion, there was a large snowstorm and many of the fishermen had a very difficult time coming ashore, some coming in very late.

Josh Allen was Emerson's uncle by marriage. He was so knowledgeable about lobster fishing that he could take any person's job if that individual was sick or had to miss a day.

Emerson died in 1973 but Jean remained active in the business and still lives next to the store in Port Howe.

Northport

Walter Brander

Walter Brander's father, Thomas, was a lobsterman who fished for Alf Trenholm. Walter's two older brothers fished predominately for "Little George" Allen, so-called to differentiate him from others of the same name.

Thomas was born in 1878 and started to fish in the early 1900s. In those days, there were lots of lobsters, some even on the shore. Walter's mother, Hazel, cooked for Josh Allen at Robinson Brook in the years 1936–1938. Walter, as history shows, is part of a lobster fishing family and plans to pass his gear on to his son, Jeffery.

Walter was born in Northport in 1929. When he was 15, in 1944, he was a helper with Walter Brownell fishing the fall season out of Northport. Walter then fished a season for both Alf Trenholm and Josh Allen. In 1946, Walter went to work in sawmills. In 1949 he joined the Navy.

After twenty-five years in the Navy, Walter returned to Northport and bought inexpensive gear from Oakley Brownell in 1974 at a time when the lobsters were very scarce. He sold his catch to Little George and then to Phil Elliott, who bought out Little George. Phil still runs a business out of Wallace.

In 1983, the lobsters began picking up and by 1985 there were all kinds of them. Walter's wife fished for him for ten years between 1985 and 1995. They saw as many as fifty lobsters in one trap during those good times but in 2003, his best total haul on any given day was ninety pounds!

There used to be nine fishermen out of Northport and now there are only seven. Everyone gets along well. There were always a few fishermen out of River Phillip, including Reg and Vaughn Pauley, as well as Chandler Angus. There were a few out of Tidnish at one time but not now.

There were never any factories along Northport's sandy beaches as there were elsewhere on the Gulf, such as at Wallace and Malagash. Northport benefited from a sheltered harbour at the mouth of the Shinimcas River and its lobster factories were built along these waters. In the late 1930s and early 1940s there were two or three lobster factories, all near the wharf. John O'Black had one about a half mile to the east of the harbour and Alf Trenholm had another about 500 yards to the north of the river's mouth. There was another as well but Walter can't remember who had it.

Many changes have taken place since Walter first started to fish:

When Walter worked in 1944	Today
Gas engines	Diesel engines
38-foot boats, light & narrow	40 to 42-foot boats, much wider
No cabins	Sophisticated cabins
No toilets	Toilets
Few women workers	Quite a few women workers
Grapple or sounding leads	Echo sounders
No radios	Radios and/or cell phones
Wooden traps	Wire traps
Manila rope	Polypropylene rope

Rope and pull steering	Hydraulic steering
"Nigger Heads" for hauling	Hydraulic haulers
Company ownership of gear	Individual ownership
Lobster sizes and definition more liberal	Increased size and definition of legal lobsters
Several lobster factories	No lobster factories
Traps organized by knockers	Individual traps set

Edison Brownell

Edison was born in 1920 and started fishing in 1937 when his father, Alva Brownell, lost the double license that enabled him to fish both the fall and spring seasons. Edison fished Northport in the fall using his father's boat. His father fished the Gulf Shore, just off where the golf course is today, in the spring. Either of them could help the other out but only one could hold the license for the season.

The first boat Edison used was his father's. It was controversial to use the same boat for both seasons but it was often done. At times they fished from a thirty-two-foot pink or rounded stern boat, owned by Alf Trenholm (Hudson senior's father), a lobster packer who would then pay so much a pound for the boat's lobster catch. However, he did not pay them in cash. Rather, Alf supplied the fishermen for the season and then would set the value of each catch against the fisherman's individual supply tab.

If the value of a fisherman's total seasonal catch exceeded the value of the supplies Alf had provided, the lobster man was paid the residual in cash. But most people owed packers like Alf. Even in the wintertime, the packer would be the fishermen's banker. Everyone thought Alf and other packers were honest brokers. In 1938, Alva and Edison sold their catches to Alf but only saw cash for the lobsters when the size of the catches increased. Later Edison was offered more money than Alf was prepared to pay so he moved his business to another packer on a cash basis.

Alf had a factory in Northport located on the Sand Point, which is now mostly underwater. Lorne Chapman owned two fish factories on the point and two more on the other side of the river, although these were strictly for smoking. Herring were brought from as far away as the Magdalene Islands for processing.

When Edison came back from the War, he operated a store and then decided to get back into fishing. In 1952, he bought his first boat. Edison's first helper was Bob Miller from Amherst but he left early because they didn't get along. The boat came from the Kennedy's in Wallace and, unlike the pink boat, had a "V" stern.

Edison has seen substantial change in the boats since 1937. At that time, for example, there was no fathom meter. You had to toss a weight overboard, either a sounding iron or a grappling hook, to determine what type of ocean bottom you had.

By 1952 there were three packers buying lobsters from Northport as well as one smack boat[1] or independent purchaser who bought lobsters directly from the buyer. The dominant buyers at that time were: Josh Allen, Pugwash (Murray Fisher bought for him), Ken Kennedy and Little George Allen, Josh's brother and the owner of the smack boat which bought from the fishermen on the water.

Edison sold his lobsters primarily to Josh but would sell two or three crates to Little George's smack boat, which paid nominally more. For quite a few years in the 1950s and 1960s Wilson Shatford, who had a factory near Hubbard's, bought Edison's market lobsters. Wilson paid more money than others.

Josh dealt with all aspects of the lobster fisherman's employ and was considered honest. On one occasion, while Edison was still operating the store, Josh tried to sell him some rope at prices higher than wholesale: the price Edison would pay at the store. Josh explained that he had to charge a higher price to make a living but Edison didn't buy.

Fishing was very poor in the 1970s. Edison did not fish during those years but worked as a supervisor on the roads. In 1978, the government changed and he lost his job so he returned to fishing.

The lobster trade had its ups and downs. Fishermen figured there was a fourteen-year lobster cycle: the lobster catches would improve progressively peaking in the seventh year, and then decline, for another seven years. For many years, there was no official limit on the number of traps fishermen could set. Therefore, to offset smaller per trap catches in periods of decline they would set out large numbers.

Edison generally caught between 12,000 and 15,000 pounds a season. The best season was 1985 in which he caught 52,000 pounds. The price wasn't like it is today but it was still pretty good.

One time Edison was caught fishing on Sand Reef, eleven miles from port just as a hurricane was rising. There was no housing on the boat and he had a very miserable time before he got into the wharf.

Edison stopped fishing in 1987 at the age of sixty-seven. He always fished his gear out of Northport. His son fishes mackerel for bait north of PEI to Antigonish. He sells in Antigonish but always brings a load home for Earl Chase, the Port Howe buyer, every week.

Note

1. Smackmen first appeared in Maine in the 1820s, a time of increased demand for lobsters from New York and Boston markets. Smackmen were named after their boats, "well smacks." These were small sailing vessels with a tank inside the boat that had holes drilled into it to allow seawater to circulate. The smacks were used to transport live lobsters over long distances.

Glossary

Grapples: Long poles with hooks on the end or just a grappling hook on a long rope (a small weighted hook with several arms for snagging traps or lines under the water). During spring season moving ice can catch buoys and drag trap lines for miles. Often fishermen used to tie poles to trap knockers and they would glide under the ice and grappling hooks or poles were used to pull the trap line in.

Hauler: A round revolving pulley operated with the engine that helped pull the trap line into the boat. The fisherman would grapple the line, wrap it around the hauler and it would do the work pulling the traps in the water up to the boat. They were shiny and were mounted on the stern of the boat. They were called a "nigger head" by most people into the 1980s. Changes in the design and political correctness changed that.

Knitting the Heads: The traps used to have hand made heads, or netting, in the openings in the wooden trap. They were made in the winter and spring by everyone in a fisherman's family, Gramma down to kids. Fishermen paid a few cents each for a head. They included two trap ends, two for the side opening hoops and the interior trap separating the "Kitchen from the Parlour"

Knockers: A grouping of traps under one buoy. Years ago when there were few trap restrictions, huge trap lines were fished. In the sixties and seventies knockers under one buoy were usually six to ten. Now with the few traps allowed and Satellite mapping systems usually only one or two traps are used per buoy.

Make and Break Engines: Old two-cylinder "putt putt" engines. They were easy to repair because they were a simple design. They also broke down easily and often.

Pinky Boat: A type of boat with two pointy ends. Not as stable as a normal square stern boat. The square stern gave a wide surface with good footing to haul traps up on to empty them. The pinky boat was easier to haul trap lines with, though. When you pull up knockers and trap lines, you work from the stern of the boat. You hook a line and you are actually pulling the boat backwards along the line. A pointy end makes it easier to pull backwards.

Smack Boats: Buyers for fresh lobster. Not locals, they would come to a fishing port and buy from boats bringing their lobsters in and take them by water to a market.

Definitions kindness of David Dewar, Curator of Wallace and Area Museum